MEGAFIRE

MEGAFIRE

THE RACE TO EXTINGUISH
A DEADLY EPIDEMIC OF FLAME

Michael Kodas

Houghton Mifflin Harcourt

Boston   New York

2017

For information about permission to reproduce selections from this book, write to trade.permissions@hmhco.com or to Permissions, Houghton Mifflin Harcourt Publishing Company, 3 Park Avenue, 19th Floor, New York, New York 10016.

www.hmhco.com

Library of Congress Cataloging-in-Publication Data is available.
ISBN 978-0-547-79208-8

Book design by Chrissy Kurpeski

Printed in the United States of America
DOC 10 9 8 7 6 5 4 3 2 1

CONTENTS

PROLOGUE

THE FOURMILE CANYON FIRE BROKE OUT on Labor Day 2010, while my wife and I were moving into a cottage in the Colorado Chautauqua, a National Historic Landmark and park overlooking Boulder.

Our home for the year at the foot of the Flatirons — pinnacles of red rock that define the city's skyline — carries a rugged, rustic status. The historic dining hall and auditorium 100 yards below our cottage hosted Teddy Roosevelt a century before we moved to the park and bands such as Los Lobos after we arrived. At night we listened to the concerts from the easy chairs on the cottage's front porch while staring at the lights of the city sprawling below a forest of stars. In the morning we looked out on hot air balloons ascending in the sunrise. On game days we could see the crowds and hear the cheers in the stands of the University of Colorado's football stadium. The panorama of dark mountain forests descending to the glowing city highlighted what made Boulder County one of the fastest-growing counties in the West.

The cottage's interior was as rough-hewn as the Rockies. Ancient knob-and-tube electrical wiring drooped from the ceilings. A cockeyed brass chandelier seemed like it might burst into flame if you looked at it wrong. The windows sat off kilter in their sills. Stinkbugs and box elder beetles didn't even have to slow down to fly inside through the gaps. Sunlight beamed between the slats into the dim, dusty interior.

Between the wiring, the wood, and the trees surrounding the cabin, our new life seemed almost designed to burn, but the charm of the location all but blinded us to that. We raved about the delicious apples that fell from the tree overhanging the front gardens, but never discussed the fire hazard posed by the pine needles that piled up in the corners of the roof and the century-old clapboards.

As we unloaded our boxes, we heard the first planes and choppers buzz over our heads on their way to the wildfire that was already devouring homes. We listened to a police scanner on a laptop and wondered how quickly the ancient wooden cabin would ignite if a firebrand landed on it. I quit unpacking and climbed to the top of the First Flatiron. From there, more than 1,000 feet above our new home, I could see the column of smoke rising five miles away.

On the shoulder of Flagstaff Mountain, farther west, I joined nearly 100 other Boulder residents looking out on the flames spreading over Sunshine Canyon and Sugarloaf Mountain. The vista was similar to the one I had over the city from my porch, but the scene wasn't nearly so idyllic. Instead of balloons in the sunrise, we watched air tankers paint red crescents in the sky just above the flaming pines. We could see some of the more palatial homes in the canyons and watched as the fire worked its way among them. The houses didn't explode, but gradually showed a tiny bit of red around their bases or on their roofs.

"That one's going up," someone would shout as soon as they saw the slightest glow. Sometimes the houses burned fast, but usually it took 30 minutes or more for them to become fully engulfed. Despite covering dozens of wildfires, and fighting them over a summer, this was the first one I saw destroy people's homes.

The Fourmile Canyon Fire burned 169 residences and was the first of four fires in four years that would break the "most destructive" fire record in Colorado. While it never threatened Chautauqua or our cottage, it did burn away many of the ideas I had about wildfire. The houses of Fourmile Canyon would be rebuilt long before I filled those voids in my understanding.

Five years later a fire more than a thousand miles away hit even closer to home. In Lake County, California, my brother Jeff was living in a cabin similar to our onetime home in Chautauqua. During the height of the summer of 2015 the Rocky and Jerusalem Fires burned within a dozen miles of Jeff's cabin. Then, in September, the Valley Fire burned 76,067 acres through the mountains where he lived.

"We'd been seeing it all summer," he told me when I called. "I saw the smoke over here and thought, 'Oh, shit, here we go again.'"

As the blaze came toward his cabin, Jeff found that his truck wouldn't start. He was left with a tiny, two-door sports car to carry anything he wanted to save and raced away with his girlfriend's paintings and silk screens, some jewels and gold, a few tools, and two cats (one would escape his grasp while he searched for a shelter and perish in the flames).

During the following days he did what he could to help others who were evacuating—gathering pets, turning off electricity, and leading a disabled friend away from the fire. Eventually he found himself alone on a ridgeline, where he watched the panorama of fire.

"I could see for miles—almost all the way to the East Bay," Jeff, an unflappable marine veteran who served in Vietnam, said. "It was like that mountain got napalmed and rocketed all at once."

By then he was certain that his own home had burned.

"Going back out the last time was like driving out through the apocalypse," he said. "Whole neighborhoods have just vanished. They're gone."

By the time my niece and her mother, who live in nearby Cobb, gathered their pets and some valuables, the flames surrounding the town were so thick that they couldn't drive through them. They parked on a golf course, where treeless greens, water hazards, and sand traps provided a refuge in the orange night. Their home survived, but many of the ones around it, and most of nearby Middletown, burned. They couldn't return home for weeks.

The Valley Fire burned more than 1,300 homes and killed four residents,[1] one of them an acquaintance of my brother's. It also climaxed the year in which the most land on record in the United States—10 million acres—burned in wildfires.[2]

I covered my first forest fire nearly 30 years before the Valley Fire burned my brother's home. I fought fires in the Rocky Mountains a decade before the Fourmile Canyon Fire blazed outside Boulder. But the new fires were different from the ones I'd photographed, reported on, and fought years before. Some scientists and firefighters were calling the worst of them "megafires," but others bristled at the term's sensationalism and lack of scientific precision.

During the five years between the fire outside my hometown, which began the series of record-breaking fires in Colorado, and the one that burned down my brother's cabin and topped off a year in which the most acreage in the nation burned, I tried to learn what was driving fires to be so much larger, faster, hotter, and more destructive. It seemed like that question would be easy to answer, particularly with the world's first conference studying megafires announced soon after the fire that threatened Boulder.

But just defining "megafire" proved more difficult than I'd anticipated.

THE CONFERENCE WASN'T IN A VAST, western forest under a scorching summer sun, but in a hotel auditorium in Tallahassee, Florida, a week before Thanksgiving 2011. The opening speaker was the first person I had contacted about megafires — the man who had coined the term. Jerry Williams was once the national director of fire and aviation management for the U.S. Forest Service, and thus the top wildland firefighter in the United States. He was the first to warn of an explosion of wildfires that would see the amount of land burning in the United States triple since the 1970s. (During the conference scientists predicted that in some years U.S. fires would burn almost seven times the amount of land burned in an average year in the 1970s.)

"My first experience with a really unimaginable fire . . . was in Northern California late in August in 1987," Williams told me when I first called him. That "fire siege" was made up of more than 1,000 blazes that brought every available wildland firefighter in the nation to California and led the governor to declare a state of emergency in 22 counties.

"I remember saying, 'Jesus, we will never see anything like that again,'" he told me. "And the next year we saw Yellowstone."

After he retired from leading the U.S. wildfire fight, Williams led a team of international researchers that studied eight megafires around the world for the United Nations.[3]

"We're seeing . . . a new type of fire . . . in the U.S., Russia, Australia, Greece, South Africa," he told the assembled scientists in the opening presentation in Florida. "It seems like every year we see a 'worst one.' And the next year we see a worse one yet. They're unbounded."

After leaving the Forest Service, Williams became increasingly critical and outspoken about the war on wildfire that he once led. Policies promoted to protect people, homes, forests, soil, and even our air and water often backfired. Many blazes we extinguished simply set the stage for bigger and badder fires in the future.

"How many of those fires were the result [of], were predisposed by, a land management decision made years earlier?" Williams asked.

Continuing to fight the fires as we have, he said, was a dead end. Yet we continued to wage an increasingly costly war in which every battle we won seemed to bring a greater loss in the future.

"We're trapped by the myths of our own success," he said. "Sometimes I think this is almost a religious issue, that we can somehow dominate it."

The three days following Williams's presentation were an apocalyptic travelogue as scientists described megafires' impacts on the atmosphere and soil, insects and flowers, economics and culture. Toward the close of the conference I sat down with 10 researchers from China, France, Portugal, Australia, Canada, and the United States in a discussion led by Dan Binkley, a forestry and ecology professor at Colorado State University. Our charge was simply to come up with a definition of "megafire." But despite the hundreds of years of cumulative research the group had done on the topic, none of us were clear about what we were actually talking about.

"The term 'mega' must be used to describe the top level of fire intensity," one researcher insisted.

"It's impossible to escape a megafire," another threw out.

"They're impossible to control."

"Israel [the Mount Carmel Fire of 2010] wasn't a megafire."

"I don't see any reason that Australia [Black Saturday in 2009] wouldn't be a megafire."

"Africa burns most," one researcher noted.

"You're looking at the amount of acres burned," Dan said. "But you can look at the number of lives lost."

"In 2003, in Portugal, the fires had political impacts," Célia Gouveia, of the University of Lisbon, said.

"The socioeconomic impact would have to be part of the definition."

"The ecological impacts should be considered."

"The definition should have multiple criteria: property loss, air quality, wildlife, steady-state ecosystem change, species extinction . . ."

The conference program, and posters around the hotel, showed a lack of agreement even about what the word "megafire" should look like. Some wrote it as one word, others as two, and the conference organizers hyphenated it.

"Is there a term better than 'mega'?" Dan asked.

"High-impact."

"Catastrophic."

"Uber," one Canadian scientist said, chuckling.

"Hellfire," Dan threw out, to guffaws.

The group fell into the weeds of fire science, arguing about whether "burn intensity" or "spread velocity" is a better measure of a fire's meganess.

"There's an intensity there that's sort of beyond our experience or comprehension," Dan said. "Jerry Williams and his colleagues saw fires with behaviors they had never encountered before."

Snow was falling in much of the West, but the following day, while we were visiting the Tall Timbers experimental forest near Tallahassee, a fire burned into the suburbs of Reno, Nevada. "My house is threatened by a megafire," Tim Brown, a scientist from the Desert Research Institute, announced with a shocked smile after calling some of his graduate students to check on his property.

The phenomenon seemed to be reinventing itself and closing in on us, even as we struggled to define it.

I returned from Tallahassee to find my home state exploding with its most destructive wildfire season in history — one that was doubly devastating when I took into account that one year burned right into the next, without the usual interruption of winter. Eventually that would drive me to consider another definition of "megafire" — duration. How would a year-round fire season add to the impacts that the researchers I'd sat with had listed?

Four years after the conference the U.S. Forest Service would define a "megafire" as a wildfire that burns at least 100,000 acres. By then, I had developed more nuanced measures. In charred forests around the world and in the endless season of flame that plagued my own backyard, I found four categories of drivers behind the flames.

- Our use and management of forests left many of them overloaded with far more vegetation than naturally grew in them, as well as invasive, introduced, and exotic species that disrupted the historical cycle of fire.
- Booming development into flammable landscapes provided another fuel load in the form of homes and infrastructure, filled forests with human-produced sparks and heat, and complicated firefighting and forest management.
- The warming and drying climate primed many wildlands to burn and expanded fire seasons by months, often into times when few resources are available to deal with them. And humans had also expanded the fire season with sparks from power lines, vehicles, campfires, and firearms that ignited fires in months in which there was no lightning to start natural wildfires.
- Political and economic decisions intended to deal with wildfires drove the flames as often as they snuffed them.

I also came to see that despite the size and ferocity of the last decade's fires, the biggest and baddest of them are still to come. Since the turn of the millennium four different years saw more than 9 million acres burn — record amounts of land that were unthinkable just a few decades ago. Then, in 2015, wildfires spread over more than 10 million acres of U.S. forests. Fire scientists anticipate that within a few years, 12 to 15 million acres a year will burn, and U.S. Forest Service researchers warn that by mid-century that number could reach 20 million — an area nearly the size of Maine.

As I chased fires across Colorado and around the world, each conflagration illuminated at least one of the drivers of the world's crisis of fire. But all of them came to play on Yarnell Hill, Arizona, where a small blaze killed the greatest number of professional wildland firefighters in U.S. history. Whereas my first trip to research what was bringing an exponential increase in wildfires to the planet had left me with more questions than answers, my last stop showed what is at stake if we fail to answer them.

PART I

EVERYBODY'S HOMETOWN

1

YARNELL HILL

Yarnell, Arizona — July 1, 2013

EARLY ON THE FIRST MORNING of Prescott, Arizona's Frontier Days — the oldest rodeo in the world — three white pickup trucks from the city's fire department drove across a ranch 35 miles southwest of the community filling with cowboys and cantankerous bulls. The trucks drove west on a track bulldozed overnight through what firefighters call a "moonscape" — land burned so black and bare that the path through it seemed to have an amber glow. The fire that had burned the Weaver Mountains the day before had been so hot it cracked the granite boulders strewn about the canyon above them. Where they stopped, only a few black and skeletal sticks remained of what had been an eight-foot-tall bramble of prickly manzanita, catclaw, mesquite, and scrub oak. A single surviving cactus rose from the ash where a dozen firefighters stepped out of the trucks.

Most of them had known one another for years. Some had watched their children grow up together or had gone to school with one another. One taught a firefighting class at Prescott High School, where several of the men who drew the trucks to the ranch had learned the craft. All of the arriving firefighters had studied wildfire with one of the men waiting for them in the canyon.

Some might have chosen to wear more formal attire for their somber reunion, but although there was barely a wisp of smoke in the air, pockets of chaparral were still burning in the mountains above them and the rubble of the town of Yarnell smoldered less than a mile to the east, so their matching fire-resistant uniforms — green pants and yellow shirts — were not only a sign of respect but a requirement to stand in the burn zone.

Nearby, 19 orange bags lay in two rows on the ground. Beyond them, scraps of charred fabric and foil peppered a 30-by-24-foot oval of ash and cinders.[1] The previous afternoon 19 hotshots — the Special Forces of forest firefighting — had crawled under the sleeping bag–sized fire shelters from which the silvery fabric came. After the blaze that had trapped them in the canyon passed over, some shreds of their shelters looked like charred chewing gum wrappers.

Faint rectangles in the shape of the fire shelters marked the charred ground, like impressions left behind by tents pitched before a rainstorm. Boot prints in the burnt ground carefully orbited the door-shaped marks and weaved between them. Other prints buried beneath the ash came from feet that had frantically raced around the area before it was overrun by the waves of flame.

As the men from the trucks added their own boot prints to the scene, they looked for clues of what had happened there. Every piece of evidence they could see pointed to a rushed and frantic deployment. Three melted and scorched chainsaws lay 20 to 40 feet southeast of the site, and another was found within the perimeter, as were packs with flares known as fusees; fuel cans, most of which had exploded; and hand tools such as axes. Some of the gear was found underneath the firefighters, despite their training to bring only water bottles and radios into their shelters and to throw everything else as far as possible from their deployment site to avoid adding their own fuel to the fire.[2]

Most of the tools' wooden handles were charred, some burned all the way to ash. Fiberglass had disintegrated into threads.

There were at least eight sawed stumps east of the shelters, but the farthest was just 45 feet from the deployment zone, a fraction of the distance the firefighters would have wanted to clear of vegetation to keep the flames away from their shelters. Fusees with their caps off showed they had tried to burn away other brush.

When the first paramedic landed in a helicopter after the long search for the missing hotshots, he was heartened to hear a voice coming from inside one of the shelters. When he went to investigate, he discovered that the source of the chatter was one of the hotshots' radios. There were no survivors.

He found just seven of the firefighters fully inside their shelters.[3] The others either hadn't climbed all the way inside before the flames arrived or had tried to get out to escape the heat when the fire was on them, or their shelters had been blasted off them. Only 13 had their feet aimed at the fire, and just 10 had gloves on to protect their hands from the heat while they held down the shelters, despite their training to do both those things. Five were lying on their backs rather than their stomachs as an effective shelter deployment requires. None of them, however, appeared to have panicked and tried to outrun the fire. They may not have had time to complete everything they were trained to do as the disaster unfolded, but they held fast to the most important part.

Winds in excess of 50 miles per hour had pushed 70-foot flames nearly horizontally into the canyon that had trapped the hotshots. Investigators determined that the fire front had traveled about 100 yards in 19 seconds. A well-trained firefighter takes about 20 seconds to deploy a shelter.[4]

All the hotshots had holes burned through their flame-retardant clothes, which would have required temperatures of about 825 degrees Fahrenheit — nearly three times what a human can survive in any prolonged exposure.

Extreme radiant and convective heat transferred so much energy that there was no difference in the temperature between the front and lee sides of the shelters, and little between the front and back of the deployment site. Fire shelters can reflect 95 percent of radiant heat, but the foil begins delaminating from the cloth around 500 degrees Fahrenheit. Once that happens, their ability to protect their inhabitants plummets.[5] The foil in the shelters melts at 1220 degrees, the fiberglass breaks down at around 1500 degrees, and the silica cloth turns brittle at 2000 degrees. At those temperatures, the condition of the shelters was likely irrelevant — most of the firefighters would have died from a single breath of superheated air.

TWO OF THE FIREFIGHTERS who came to retrieve the hotshots' bodies draped the orange bags with American flags. Others quietly walked among them whispering prayers or silent good-byes. Then the team regrouped and rotated through the tasks they'd volunteered for. Eight split into an honor guard leading to the back of a pickup, four on either side. The remaining four lifted a bag by its corners and carried it to the truck. As they approached, the other firefighters snapped to a salute. Once a bag was secured in the truck's bed, the honor guard stood at ease, the men wiped their eyes, and the teams rotated.

The bags were marked by numbers, rather than names, as many of the bodies inside them were burned beyond recognition. Members of the honor guard were thankful they wouldn't know which of their friends they were carrying until the investigators' reports were released two months later. The radio the paramedic had heard the night before continued transmitting voices that repeatedly startled the crew into hopeful glances toward the deployment zone.

Once the last corpse was loaded, Darrell Willis opened the Bible he carried in his truck to Psalm 23. A former chief of the Prescott Fire Department, Willis had come out of retirement to take on the task of protecting the city from forest fires as the department's Wildland Division chief. Under his leadership the Granite Mountain Hotshots had formed — a crew of 20 elite forest firefighters that in 2013 was the only one of 109 hotshot crews in the country to be part of a *city* fire department. Only one member of his crew was still alive when Willis read the psalm in the charred canyon.

Yea, though I walk through the valley of the shadow of death, I will fear no evil.

Danny Parker, a firefighter with the Chino Valley department northeast of Prescott, fell to his knees as Willis read.

"I'm a broken man," he said later. "Just weak and broken and humble."

Danny asked if he could pray with the men. His voice quivered and cracked, and the other firefighters wondered how he could possibly deliver even a short invocation, given that he had lost more than anyone else at the scene.

WHEN HE WAS A CHILD in Prescott, Danny sat on his back porch with his family watching the 1972 Battle Fire as it licked over the mountains where they often hunted and picked apples. It grew into the largest blaze in Arizona history to that date. He saw the Castle Fire break that record seven years later, and the Dude Fire that grew nearly as big and killed six firefighters near Payson in 1990. After stints in the U.S. Navy and working construction, he joined the Chino Valley Fire Department.

Danny and his wife raised four children, but his son Wade followed closest in his footsteps. A photo of Wade as a child shows him wearing a firefighting costume and hosing down a small bonfire in the family's backyard. For his school's science fair, he built a model demonstrating the chimney effect on a house fire.

Wade excelled at every sport he tried, but baseball and bow hunting were his passions. Danny made all of his children's bows and gave Wade his first one when the boy was three years old, then made him others as he grew bigger and stronger. The weapons are works of art — made of woods such as bird's-eye maple, African wenge, and Brazilian yellow-heart, and adorned with elk horn bow tips. Father and son would text each other photos of the tight patterns of their arrows on targets. Danny also coached several of his son's baseball teams and would get flack from family and friends for not highlighting Wade's talents more, not that his son needed it. A grand slam that Wade hit in high school helped him secure a baseball scholarship, but his classic ballplayer's good looks and polite demeanor — sandy brown hair, wide boys-of-summer smile, and Sunday-school manners — probably didn't hurt.

"We always thought he'd be famous for getting to the big leagues," his older sister, Amber, said at a church service 10 days after Danny Parker's prayer in the burn zone. But after a year in college, Wade gave up his scholarship and moved back into the family home.

"He loved baseball, but he said he realized it was 'time to become a man,'" Danny said. "He wanted to come home, pursue becoming a firefighter, get married, and have a family."

Danny advised him to get a job on a Type 2 hand crew, the common entry into the world of fighting forest fires. "But, no, he wanted to be a hotshot," Danny said.

Nobody was surprised when Wade was hired onto Prescott's elite Granite Mountain Interagency Hotshot Crew, or when he was named the crew's Rookie of the Year in 2012.

Father and son often hugged as they left their house for their respective firehouses, then chatted about their workdays when they were back at home. "Never lose your respect for the fire," Danny told his son. "That's when bad things happen. The power of the forces on this earth, they don't care who you are."

When they hunted together, Danny would point out fire hazards to his son, such as ravines that could trap a firefighter in a natural chimney, or clouds on the horizon that signaled weather that could turn a fire or blow it up. Thunder could make rutting elk bugle and warned of new fires ignited by lightning. Wade, like his father, worked on the side as a hunting guide, and developed a good eye for vegetation and habitat. He recognized that fire was just as often a benefit to the forests where he worked as a hazard. He told his father about areas he'd like to go back to with a bow when he didn't have to be there carrying a chainsaw.

"He didn't talk about the fires hardly ever. He talked about how beautiful the mountains were," Danny said. "About this waterfall and how they bathed under it because they hadn't had a shower in a long time."

A week before the Yarnell Hill Fire, the Granite Mountain Hotshots worked on the Doce Fire, a 6,700-acre blaze just outside Prescott — exactly the kind of threat to the city the unit had been formed to stand up against. They got something of a hero's welcome afterward for helping to protect more than 400 homes that the fire had threatened, and for saving the nation's second-largest alligator juniper tree, which was located on Granite Mountain.

But much of their namesake peak had burned, and Wade confided to his father that the crew was as humbled as they were proud. "They got teased for letting their mountain burn," Danny said.

Some would see the scorching of the mountain the hotshots were named after as an omen of what happened to them just weeks later. But before that, Danny saw something more concrete to worry about in the Doce Fire.

He was working outside on a day off from firefighting when he noticed a wisp of smoke over the mountain. Just a few hours later, the Doce Fire was racing toward the homes in the Williamson Valley. "From 50 to 5,000 acres in just three or four hours? Its rate of spread was absolutely incredible," Danny said. "It was kind of a foretelling of what the fire behavior was going to be like."

In the 25 years that Danny had been fighting fires, he'd noticed a change, and he told his son about it. "If you go back 30 years and watch bull riding, there were a few bulls that bucked incredibly well," he said. "Then they started breeding these bulls to buck. Now every bull that comes out of the bucking chutes is like the better ones from 30 years ago. The same goes with fires. We've always had volatile fires during volatile times. But now almost all of them are that volatile."

Danny was happy that Wade was working with men whom he knew and respected — and, more important, who he knew respected fire. Eric Marsh, the superintendent of the crew, had been eating, sleeping, and dreaming wildfire for decades. He and his wife at the time had started the Arizona Wildfire Academy in their mobile home and turned it into one of the largest and most respected schools for forest firefighters in the country. Jesse Steed, the captain of the crew, was a former gunner in the U.S. Marine Corps who had no problem hugging his men, telling them that he loved them, and then leading them through workouts that left them puking along the trail.

Wade patterned his work after some of the previous Rookies of the Year — Andrew Ashcraft and Robert Caldwell, both family men. A few weeks before the Yarnell Hill Fire he was promoted to "sawyer," responsible for running a chainsaw, and proved as skilled with it as he was with a bow or baseball bat.

Danny was also pleased that his son's fire crew fit so well with Wade's reverence of family and God. When his younger brother, who dreamed of being a bull rider in rodeos, turned 18, Wade showed up with several of his crewmates and their girlfriends at the small restaurant where the family was celebrating, bearing the gift of a $500 pair of chaps. In 2012, while visiting Disneyland, he proposed to Alicia Owens, his girlfriend since he was 16. Their wedding was scheduled for October 2013.

Wade was in his second season with Granite Mountain when the summer of 2013 exploded with fire. Chris Hunter, a high school teacher he'd stayed in touch with, sent him a worried note. He texted her a picture of himself on a fire. "Miss Hunter," he wrote, "I'm a hotshot. I'm good."[6]

• • •

DANNY PARKER WAS THE FIRST to volunteer to retrieve the bodies of his son and the rest of the Granite Mountain Hotshots — the only family member of the fallen firefighters to join the recovery team. At the site where the hotshots made their final stand by lying on the ground, he finished his prayer and somehow rose to his feet.

"All I could do is thank him for 22 years," he told me.

Then someone looked up at the mountain and pointed out two cracks forming a cross on the face of a huge granite boulder.

"All 19 of those boys were at the foot of the cross," he said.

As they left the scene, the recovery team placed a more humble marker on the site — a black T-shirt with the Granite Mountain Hotshots' logo, which they hung on the lone surviving cactus.

Danny knew that each of the men who perished on Yarnell Hill had a story as compelling as his son's. Every one of them was loved in the tight-knit clan that grew up around Arizona's oldest fire department. Each of them had his own family now quaking with grief. And Danny also knew that across the country and around the world, other communities were facing mass fatalities and the destruction of hundreds of homes in wildfires.

In the following months Danny would learn that Wade was the closest of the hotshots to the cross in the stone and that he had carried the bag with his son's remains. He took some comfort in that knowledge.

But the most urgent question he and other firefighters around the nation were struggling with was left unanswered by the debris strewn at the deployment site and the investigations that followed the fire.

An hour before they crawled under their fire shelters, the Granite Mountain Hotshots were standing in what firefighters call "the black" — where most of the vegetation had already been burned away, creating a haven from the fire.[7] They could see the inferno exploding below them. Yet they descended into a box canyon that any firefighter would recognize as a death trap — a natural chimney so choked with brush, they had to hack their way through it, and so oily it burned as if soaked in gasoline. What led Wade Parker and his crew to leave the safety zone where they could have easily survived the fire they could see boiling below them and descend into the canyon where they were most vulnerable to it?

The answer to that question lies as much in the century of policy decisions and the decadelong explosion of fires that led up to the disaster as it does in the decisions the hotshots made in the minutes before their deaths.

2

FUSES AND BOMBS

WITH NO WITNESSES to their final moments alive, mystery will always shroud the deaths of the Granite Mountain Hotshots. But other questions of what led to their annihilation have answers that are increasingly clear.

For years scientists, firefighters, and foresters have warned that disasters like the one on Yarnell Hill are all but inevitable and likely to recur with growing frequency and devastation. Their causes have been well-known and on the rise for decades.

During the century leading up to the tragedy, U.S. firefighters successfully extinguished about 99 percent of the wildfires that ignited in the United States, and that interruption of the natural fire cycle led many forests to grow unnaturally thick with timber, brush, grasses, and other fuels.[1] Poorly managed timber harvests,[2] overgrazing, forest pests, invasive species, and ill-conceived timber plantations are blamed for further overloading landscapes with tinder.

Climatic changes, driven to no small degree by human emissions of greenhouse gases, brought droughts and heat waves to many formerly moist woodlands. Mountain snowpack melts off weeks, and sometimes months, earlier than it did during the previous century, leaving forests dry enough to burn at times of the year when they used to be filled with streams of snowmelt. In other areas, warming trends brought pulses of precipitation that fed growth spurts of grasses and other "fine fuels" that burned intensely when arid conditions returned. According to the U.S. Department of Agriculture, the warming and drying climate has expanded fire seasons in the western United States by 78 days between 1970 and 2015.[3] Human activity on the ground has also expanded the length and range of the fire season by bringing sparks and flames to months and landscapes that have no lightning to ignite them naturally. In some woodlands that are usually blanketed with snow in winter, wildfires are now a year-round threat.

Each year homes spread deeper into these increasingly flammable landscapes. A 2015 study by the U.S. Forest Service reported that about one in three U.S. residences — 44 million homes — are in the wildland-urban interface, or WUI, where development abuts fire-prone open space.[4] Research at Colorado State University shows that between 1970 and 2000, there was a more than 50 percent increase in the number of homes beside flammable forests, scrub, and grasslands, and two-thirds of those houses were adjacent to wildland prone to high-severity fires.[5] By the 2010s, according to research by Headwaters Economics, the number of homes burned each year in U.S. wildfires was more than seven times higher than in the 1970s.[6] Records at the National Interagency Fire Center show that average annual wildland firefighter deaths doubled between the 1970s and the 2000s.

Budgets are burning as well. In 1995, 16 percent of the Forest Service's budget was spent on wildfires. Twenty years later wildfires consumed 52 percent of that funding. During that period U.S. government spending to fight wildfires increased from around $300 million to $1.5 billion annually.[7] In the decade leading up to the 2013 disaster on Yarnell Hill, the government's total tab for preparing for, fighting, and recovering from wildfires averaged more than $3 billion a year. That average was less than $1 billion in the 1990s.[8]

EACH OF THESE FACTORS came to bear on the Granite Mountain Hotshots, both during the months they traveled around the nation fighting enormous fires on federal land and in their final moments, when they were overrun by a small blaze in their own backyard. Their home turf, in fact, is among the areas of the planet with the most rapidly increasing wildfire hazard.

At the time of the Yarnell Hill disaster, Arizona, according to a report from Climate Central, was enduring a combination of warming and drying faster than any other state in the nation.[9] After a century of extinguishing every wildfire it could, researchers found that some Arizona forests that historically had fewer than 20 trees per acre now have more than 800 — 40 times their natural density of fuel, a condition endemic to ponderosa pine forests of the Southwest.[10]

Development is spreading into those ever more flammable forests as fast as in any other part of the nation. According to an analysis by the *Arizona Republic*, the pace of home building in the state's fire-prone landscapes jumped by 91 percent between the 1980s and the 2000s.[11] While those homes are increasingly palatial, few of them meet the standards of Firewise programs from the National Fire Protection Association or Community Wildfire Protection Plans from the U.S. government's Forest and Rangelands initiative. Some homeowners who bristle at any government telling them how to manage their land also expect it to provide protection for their property regardless of the hazard they create, thus magnifying the risk.

Officials often promote strategies to manage wildfire based on economic or political considerations rather than science. The "fire-industrial complex" that developed to combat the blazes is today worth billions of dollars and is often driven more by business interests than ecological or safety considerations.

Arizona is not unique in its increasing exposure to the hazard of wildfire, and the tragedy on Yarnell Hill is but one of a series of disastrous fires that are increasing in frequency, intensity, and destructiveness on every forested continent. While the deaths of the Granite Mountain Hotshots would draw more international media attention than any previous forest firefighting disaster in the United States, mass-fatality wildfires are also increasingly occurring overseas, often with far greater death tolls than that of Yarnell Hill.

CONFLAGRATIONS ELSEWHERE ON THE PLANET are harbingers of what's to come, for both America's forests and its firefighters, but they are also present challenges to our nation today. During the past decade the United States has sent crews and equipment to every corner of the globe to fight exploding wildfires.

In the Australian Alps, fires on February 7, 2009, killed 173 people and overran entire towns. In Marysville, most of which burned, residents had to pull the town's fire chief away from his own burning home while his wife and son died in the flames inside.[12] Sixty U.S. firefighters flew to Australia on just 48 hours' notice to help battle the blazes. Until that day Australia's McArthur Forest Fire Danger Index spanned a scale of 0 to 100, but at Melbourne Airport on "Black Saturday" the scale hit 165 after a month of the warmest temperatures in 154 years and record low rainfall. The day the fire ignited, the airport recorded its highest temperature ever. After the blaze Australia expanded its fire danger scale to account for rising temperatures and decreasing moisture. Bushfire danger there has hit the new level of "Catastrophic — Code Red" repeatedly since it was adopted.

On Mount Carmel, where in the Bible Elijah brought down fire from heaven to prove the power of his God, a 2010 wildfire killed 44 people, including an Israeli police commissioner, the police chief of Haifa, 36 prison guards, and a 16-year-old volunteer firefighter. The nation has suffered tens of thousands of wildfires in the past century, but recorded virtually no natural ones prior to that, an indication of the man-made nature of Israel's fire crisis. Since Israeli independence in 1948, the Jewish National Fund, which purchased the land for the founding of Israel, has promoted planting trees in the nation as the most Zionist of acts. Jews from around the world have contributed to the planting of forests, leading the nation that once had just 1 percent of its countryside wooded to now having 8 percent covered with forests.[13] But as the pines in monocultures matured and the warming climate dried the forests, the JNF's trees have fueled a series of increasingly destructive wildfires. Israelis have taken to calling the planted pines "firebombs" and their cones "hand grenades." Ten nations sent aircraft and firefighters to Israel to help fight the blaze on Mount Carmel. One of those nations, the United States, sent C-130s loaded with fire suppressant and retardant from German airfields and funding from Washington, D.C. The "Supertanker," a U.S. 747 converted to drop some 20,000 gallons of fire retardant, flew from California to Mount Carmel, where it made two drops on the fire before turning around and flying home. Although the Supertanker was rarely used in the United States, an updated version of the 747 returned to Israel six years later when a series of arsons ignited forests across the country and drove more than 10,000 people to evacuate Haifa.

In 2012 Spain was overrun by its largest wildfires in two decades, including one that trapped some 150 tourists in their cars. Dozens stumbled or slid down the rocky hillside below the flames, arriving at the seaside with burns, bruises, and broken bones. A French family of five, lost in the smoke and chaos, came to the edge of a cliff instead of the beach. Pascal Couton died instantly when he fell onto rocks in the sea below. His 14-year-old daughter, Océane, drowned. In all, nearly 600 square miles of Spain burned, killing firefighters, a forester, and a farmer, and driving residents to shelter in caves. Blazes in the Canary Islands forced 5,000 to flee, with nearly a thousand of them boarding ferries to escape.

In 2015 the smoke from some 100,000 intentionally set Indonesian fires spread over 3,100 miles and drove more than half a million people to seek medical care. The haze from the fires forced the closure of airports and schools as far away as Thailand. Indonesia prepared warships to evacuate villagers trapped in the smoke for months, but not before more than 20, most of them children, choked to death in the toxic orange fog. The United States sent more than two metric tons of firefighting equipment, and Russia sent jets to help battle the conflagrations. The *Guardian* newspaper described them as "almost certainly the greatest environmental disaster of the 21st century,"[14] and the nation's Meteorology, Climatology, and Geophysics Agency called them a "crime against humanity."[15] By year's end, the fires had done over $14 billion in damage — more than twice the toll of the 2004 tsunami, which killed thousands of Indonesians.[16] And their impact has been global. The scale of the burning since the 1990s has led successive Indonesian presidents to apologize for the haze that blankets Malaysia and Singapore every summer, and in bad years spreads across Southeast Asia. The fires, almost all of them intentionally started by multinational corporations and slash-and-burn farmers, often burn underground into the spongy peat that lies beneath many of Indonesia's rain forests and holds up to 10 times more carbon than the trees. On some 40 days during 2015 the Indonesian blazes released more climate-warming CO_2 than the entire U.S. economy,[17] thus warming the global climate and driving increases in wildfires around the planet.

Six months after the Indonesian fires subsided, hundreds of blazes ignited in western Canada as early-spring temperatures rose as high as

90 degrees — as much as 40 degrees above normal and roughly as hot as Miami.[18] Snowpack fell as low as 85 percent below average. The worst of the fires destroyed some 2,400 buildings in Fort McMurray, the Alberta city developed to service the tar sands, which produce oil with the greatest impact on the climate. More than 100,000 people in the city deep in the northern boreal forest and the camps serving the oil fields fled on the only highway leading south. Much of the population was evacuated more than once as the fire that covered nearly 400 square miles when it entered the city grew to more than five times that size — nearly as big as Delaware. Firefighters from as far away as South Africa joined in the effort. As in Indonesia's tropical forests, the destruction of the northern pine forests is magnifying the climatic changes driving the fires. The increase in wildfires and insect infestations driven by the warming climate has turned boreal forests around the world from carbon sinks that suck planet-warming CO_2 from the atmosphere to carbon sources that are releasing more greenhouse gases than they sequester.[19] The fires also amplify the impacts of climate change on ice and oceans, as black soot in the smoke travels as far as Greenland, where it peppers glaciers, absorbs the sun's heat, and speeds the melting of ice sheets, thus raising sea levels with the flood of runoff.[20]

In Russia wildfires burned some 75 million acres of boreal forest in 2012, dwarfing the U.S. record of 10 million acres.[21] Blazes in the summer of 2010, the previous record year there, blanketed Moscow with so much smoke that it, along with the heat wave that drove the fires, doubled the normal summertime death rate in the city.[22] Similar fires are increasingly burning into cities around the world — Colorado Springs in 2012; Valparaiso, Chile, in 2014; Cape Town, South Africa, in 2015; and Fort McMurray in 2016.

In Greece the massive wildfires of 2007 killed 67 people, threatened the relics of Olympia, burned into the suburbs of Athens, and nearly brought down the nation's government. In 2009, 2010, 2013, and 2015, more fires threatened the nation's forests, cities, and governments. The fires exhibited behaviors nobody in Greece had ever seen before, baking stone ruins into piles of lime and blowing firefighting aircraft back from the flames. Firefighters saved only the most cherished antiquities and let entire villages burn. Ironically, the fires in the birthplace of democracy have paralleled the nation's elections, with most election years bringing new waves of fire and closer elections bringing more flames.[23] Wildfires have become such a scourge in Greece that when the Czech artist David Cerny sculpted a map of national stereotypes for the entrance to the European Council building in Brussels, he showed Greece as a charred wasteland.

IN THE UNITED STATES, as in the rest of the world, the flames are proving impossible to stop. A decade before the Yarnell Hill Fire overran the Granite Mountain Hotshots, Arizona and Colorado simultaneously had the largest wildfire in each state's history to that date. A year later a hunter started the largest fire in California history. The "Fire Siege of 2003" burned across Los Angeles, San Bernardino, Ventura, and San Diego Counties, killing 12 people in its first 24 hours, turning 2,232 homes to cinders, and blackening 275,000 acres of land — an area more than twice the size of the entire city of Los Angeles.

Alaska set a state record the following year when wildfires burned more than 6.38 million acres of timberland and released more greenhouse gases than did the burning of all the fuel that passed through Alaskan pipelines that year. At the other end of the nation, Texas suffered its worst wildfire ever when the 2006 East Amarillo Complex spread over a million acres of grasslands, killing 12 people and more than 4,000 cattle and horses in just 48 hours.

For those who thought such catastrophic conflagrations only burned in the West, 2007's Big Turnaround Fire was appropriately named. The largest fire in Georgia's history, it burned nearly 400,000 acres and eventually merged with the Bugaboo Scrub Fire, the largest fire on record in Florida. Nine years later dozens of fires spread across the Southeast in November and December, killing 13 people in Tennessee, where they burned over entire towns.

Utah's record-setting conflagration, the Milford Flat Fire, burned over 567 square miles in 2007. Smoke from the blaze caused highway pileups as far away as California, drove truckers to abandon their rigs, and forced the closure of 150 miles of Interstate 15. Then the flames returned to California. The Zaca and Witch Fires burned from Santa Barbara to the Mexican border, killing nine people, injuring more than 60 firefighters, and destroying 1,500 homes. Emergency personnel evacuated more people during the fires than they had during Hurricane Katrina. That number would have been even higher had the call from San Diego's city attorney to evacuate the entire municipality been heeded.

Since 2000 more than a dozen U.S. states have reported the largest wildfires in their recorded histories. Beginning with the Fourmile Canyon Fire that broke out when I was moving into my cottage in Boulder in 2010, Colorado broke its most destructive fire record four times in four years.

IN 2006 — DURING WHICH a then-record 9.8 million acres of land burned in the United States — Anthony LeRoy (Tony) Westerling, a fire and climate researcher, released a paper with colleagues from across the West showing that in the 16 years after 1986, western forests saw a fourfold increase in the number of wildfires and a sixfold increase in the amount of land they burned when compared with the previous 16 years.[24] A follow-up paper released by Westerling in 2016 showed that wildfires in mid-elevation western forests had continued to increase and that grasslands and scrublands were showing a similar trend.[25]

On average, according to the nation's "Quadrennial Fire Review 2009," wildfires burned 7.15 million acres a year between 2000 and 2008 — twice the average during the 1990s.[26] Between 2005 and 2014, three years saw more than 9 million acres burn — the most since the federal government began keeping those records half a century earlier. Then, in 2015, the area burned topped 10 million acres for the first time on record. U.S. Department of Agriculture scientists estimate that in the coming decades, as much as 20 million acres will burn annually. During the 1970s that average was just over 3 million. And there's another worrisome trend in the data. Decades ago, though far fewer acres burned, there were often just as many fires. So on average today's fires are larger than those earlier in the record.[27]

According to the U.S. Forest Service, the worst 1 percent of wildfires consume 30 percent of the nation's firefighting budget.[28] Huge, hot, and fast, these fires make up a new category of fire, exhibiting behaviors rarely seen by foresters or firefighters. The infernos, often comprising many fires attacking on multiple fronts, can launch fusillades of fire-brands miles ahead of the conflagration to ignite new blazes in unburnt forests and communities. The flames create their own weather systems, spinning tornadoes of fire into the air, filling the sky with pyrocumulus clouds that blast the ground with lightning to start new fires, and driving back firefighting aircraft with their winds. The intensity of their heat makes them impossible to fight with direct attacks.

"They cannot be controlled by any suppression resources that we have available anywhere in the world," said Kevin O'Loughlin, when he headed Melbourne's Bushfire Cooperative Research Centre at the turn of the millennium. For their physical impact on the landscape Professor Stephen J. Pyne, America's preeminent wildfire historian, calls the fires "climatic tsunamis."

One of the few successes firefighters, foresters, and scientists have had in dealing with them is to give them a name — megafires.

WHEN I MOVED TO COLORADO, I joined the thousands of migrants to the West who are both threatened by and contributing to increasingly destructive wildfires. And by the time the Granite Mountain Hotshots marched up Yarnell Hill, I knew that a fire doesn't have to be big to deserve the "mega" title.

3

PRESCOTT

Prescott, Arizona — July 1, 2013

THE FIREFIGHTERS WHO RECOVERED the hotshots' bodies didn't bring them home to Prescott, but took them instead to the medical examiner's office in Phoenix. During the ride there members of the honor guard teared up as people crowded the sides of the highway — men in their dress military uniforms; children costumed as firefighters, their mothers sobbing beside them. Some saluted or held their hands over their hearts. All of them fell silent as the caravan passed.

The wives, children, parents, and siblings of the hotshots — there were more than 200 immediate family members — would wait nearly a week for the procession of 19 white hearses that brought their bodies back to Prescott. The men who escorted the bodies to Phoenix, however, returned home that night to find firefighters from across the country and press from around the world flooding the city.

Prescott, with the city slogan "Everybody's Hometown," would normally welcome the attention. Frontier Days, the biggest yearly event in the city, was scheduled to kick off as the honor guard returned from Phoenix. In the coming days, horses, bulls, and cowboys in chaps would fill the rodeo grounds. In other years at many city celebrations, such as the upcoming 150th anniversary, revelers would don period western garb. Men in bowlers or ten-gallon hats, revolvers on their hips and their gray mustaches waxed into handlebars, escorted women with parasols, fishnet stockings, and garish dance-hall dresses. But in 2013 the first day of the world's oldest rodeo was notable only for the 19 riderless horses parading through the ring.

Instead of flocking to the rodeo grounds, reporters and photographers crowded in with weeping and stunned residents lined up against the fence that surrounded Station 7 — the humble gray shack that served as the Granite Mountain Hotshots' home base. Memorabilia hung from the chain-link fence and piled up on the ground below it gradually obscured the view of the station. Within days the entire fence would turn into a wall of framed and laminated photos of the hotshots, T-shirts and baseball caps from fire departments across the nation, banners from departments in other countries, axes, boots, toy fire trucks, firefighter figurines, 19 crosses, 19 American flags, 19 shovels, 19 baseballs, 19 Frisbees, 19 teddy bears.

Television networks set up tents in the street, and satellite trucks filled up nearby parking lots. There were three crews from Al Jazeera, the news network from Qatar.

A few miles away, inside the gymnasium at Embry-Riddle Aeronautical University, family members of the fallen firefighters sat in the front rows of the bleachers or stood on the floor in the crush of television cameras, reporters, and fire crews. Prescott firefighters and former members of the Granite Mountain Hotshots sat stone-faced at the back of the stands as leaders from the city, county, state, and Navajo Nation spoke. But at the end of the ceremony, when called into the TV lights in the packed gym, the firefighters couldn't remain stoic. Dozens of them clutched one another in a scrum of wailing grief. Faces buried into shoulders; tears dampened black T-shirts, many of which had designs noting the 10-year anniversary Prescott's wildland fire crew was celebrating when the fire overran it.

Nobody inside the university auditorium, outside Station 7, or in the ash of Yarnell had more than the barest of details about the burnover that in a matter of seconds had killed nearly a quarter of the firefighters in the oldest fire department in Arizona. Some quietly wondered who would protect the city from the hazard growing around it now that the hotshots were gone.

"We're going to have to ask ourselves a set of broader questions about how we're handling increasingly deadly and difficult firefights," President Barack Obama told the nation in an address from the White House praising the Granite Mountain Hotshots on the day of the memorial.

A YEAR LATER, DESPITE TWO INVESTIGATIONS into the tragedy, millions of dollars in fines and lawsuits, and as much media scrutiny as any wildland firefighting disaster anywhere in the world had received, the details of what led to the hotshots' deaths remained largely unknown. But the devastated Prescott Fire Department and the ruins of Yarnell, when added to the decadelong list of wildfires that the disaster climaxed, provided answers to many larger questions.

The chain of events that led to the disaster on Yarnell Hill began not on the day that the hotshots arrived on the fire line, nor with the lightning strike that ignited the blaze, but in the weeks, months, and years leading up to the tragic blaze. By the same token, the wildfires that have overwhelmed the United States since the turn of the millennium grew out of economic, political, and social decisions stretching back a century. For me, comprehending the crisis began not in a flaming forest with elite, professional firefighters, but in the grasses of the Colorado plains with members of the nation's first line of defense against wildfires — a family of volunteers. One woman among them gained some small understanding of what the Granite Mountain Hotshots endured in their final minutes.

PART II

OUR GREATEST ALLY, OUR FIERCEST FOE

4

HEARTSTRONG

Wages, Colorado — March 18, 2012

JENNIFER STRUCKMEYER WAS HOME ALONE when she heard the alarm, but she knew most of her family heard it, too. Her husband's two brothers and their parents all lived in houses around the Struckmeyers' ranch near the ghost town of Wages, on Colorado's eastern plains. Their homes were almost close enough for her to shout the news to them, but they all had fire radios, so she was certain they knew about the fire as soon as she did.

It was about one in the afternoon on March 18, the last Sunday of the winter of 2012. The one person she normally would have told about the fire — her husband, Del — wasn't within earshot. Their son Austin would turn nine the next day, so Del had taken him and their 16-year-old son, Brandon, shopping with the boys' aunt in Sterling, a town 40 miles north of the ranch.

Jen had signed on to the Wages Volunteer Fire Department less than a year earlier, joining an important but overlooked demographic in Colorado, where about half of all firefighters are volunteers and all-volunteer departments protect 70 percent of the state's area.[1] Before that, she'd sometimes needled Del, who had volunteered with the department since he was a teenager, when he raced out of the house to fight fires at all hours.

"You can't go out there," she'd tell him. "You're risking your life . . . and what for? Your family's at home!"

Then she joined him on a fire call.

"I didn't understand it until I got into it," she told him afterward. "No wonder you do it. It's fun and exciting."

The sense of community was akin to that at the dances in Holyoke, Colorado, in between her hometown of Imperial, Nebraska, and the Struckmeyer ranch. She'd met Del at one of those dances. Some of the same people who attended the hoedowns also showed up as volunteer firefighters at burning houses and car crashes to do intricate dances with trucks and hoses, sometimes lit by flames in the night.

Jen's sister-in-law, Pam, was an emergency medical technician, and Jen had decided she was drawn to medical care, too. Together they would be sisters of mercy. She'd completed her certification as an EMT just a week before the alarm sounded.

Del had his fire radio with him, but Jen knew it would take him more than an hour to get to the blaze, and grass fires need a faster response than that. So Jen would fight her first fire without her husband.

A MILE AWAY JEN'S MOTHER-IN-LAW, Bev, heard the fire call and responded as she had for decades. She prayed.

Bev's husband, Lee, still went to many of the fires that the volunteer fire department he'd founded fought. But now that he was in his 70s, he wasn't the first to arrive. Their eldest son, Damon, was on his way even before the call went out. He dropped by his folks' house to say he was going in to the firehouse after hearing radio traffic about fires on the plains. He knew it was only a matter of time before the Wages department got called in, and was still in his parents' driveway when his pager went off.

Both as a fireman and a farmer, Damon's biggest worry was the weather, which seemed more like early summer than the waning days of winter. March is usually the snowiest month in Colorado, but NOAA's National Centers for Environmental Information reported 2012's to be the driest in the state's history, drier even than 2002, which tree rings showed had the lowest Front Range stream flow since 1685.[2] The state's snowpack was already less than half of what it was normally and would melt to 1 percent of average by June.

It was also the warmest March on record. Fruit trees across the state were budding months early.[3] Ranchers wondered what to do with their livestock if the drought kept them from raising enough hay and grain to feed the cattle. The University of Illinois reported the most unfavorable growing season for corn and soybeans in 25 years.[4] Grass refused to green up on the parched ground, and what did grow was as dry as paper. The sun baked the dirt of farm fields into a powder that the winter's relentless winds blew away in dust storms. Farmers complained that they wiped more topsoil from their eyes than they could see in their fields.

"If we don't get any moisture," Damon thought, "it's going to be a short season."

For farming, but not for fire. Increasingly hot and dry weather had extended the annual wildfire season in Colorado by two months. Damon and his brothers had noticed. That morning the wind, drought, and heat prompted the National Weather Service to issue a Red Flag Warning of high wildfire danger.

"Until the last few years, we didn't know about red flags," he said. "Now we hear about 'em more and more."

Lee started the department in 1985, after a number of houses near the Struckmeyer spread burned to the ground. It took so long for distant fire departments to arrive that locals joked, "Save the foundation when you get there."

Damon had joined the force the year after its founding and was on the truck that responded to the department's first fire — a burning combine. The family manned hoses when the Radio Shack burned in Sterling, an hour to the north, and fought a grass fire so far away in Kansas it took them half a night to drive to it. In 2012 the department was made up of 19 volunteers, eight of whom belonged to the Struckmeyer family. Damon, Del, and their brother Kip often rotated through yearlong terms as fire chief, on top of working the Struckmeyer ranch.

With the family making up 40 percent of the Wages department, the Struckmeyers, like many volunteers, struggled to balance their service and their family business. One fire on the family's ranch mustered the department on Christmas night, and they snuffed another that broke out during their grandmother's funeral.

"But we had to bury Grandma first," Damon recalled.

At times it seemed the Struckmeyer brothers put as much into building the fire department as they did into running their farm. Kip studied firefighting at Morgan Community College, and when the school re-placed the extrication equipment it used to train firefighters to rescue victims from car crashes, the Struckmeyers bought the old gear. A few years later the boys picked up a five-ton six-by-six tanker truck on eBay. They spent nights building a catwalk on it and welding the fittings for the hoses and tanks together.

"We'd show up at the fire hall at five or six [in the evening] and work until nine, ten, eleven," Damon said. Jen and Pam joked that their husbands must be sharing their late nights with another woman and named the truck Leslie in her honor.

Damon's daughter, Mariah, had just turned 18 — old enough to join the department. She was waiting for the meeting that would approve her coming onto the force.

Damon, at 42, was a gregarious, slightly pudgy goofball. "You can laugh, or you can cry, or you can laugh until you cry," he quipped when describing the family's struggles.

His brother Del is built like a bulldozer — six feet two and 280 pounds. A mustache droops on either side of his mouth, which holds its tongue as often as his older brother's makes wisecracks.

When Damon dropped by his parents' house on the day of the Red Flag Warning, Darin Stuart, his nephew, was visiting. Also a member of the fire department, Darin was eager to answer the call with his uncle.

A PERIMETER OF WRECKED CARS and a log-framed sign an-nouncing "Wages Fire Department" indicated that the building Damon and Darin pulled in beside wasn't just another farm shed, but the heart of emergency response for some 30 square miles of fields and 200 miles of roads. They walked inside, leaning against winds that were gusting to 60 miles per hour, grabbed their gear, and loaded up the department's fast-attack truck — a four-by-four with a platform behind the cab where two more firefighters could strap in. A control panel on the platform ran the pumps and hoses connected to the 500-gallon water tank on the back half of the truck. Damon tossed his heavy fire-resistant bun-ker coat into a compartment in the back because it cramped him when he was driving. It would be easy enough to put it on before he got into the action. They pulled out before anyone else arrived at the firehouse.

Damon's wife, Pam, heard the same call that Jen did. They're sisters by marriage but could easily have been born that way. Both God-fearing and irreverent, they share brown hair, a boisterous sense of humor, and hardy laughs. They spent years together cleaning houses and raising kids, and were looking forward to working together at emergencies. Pam, a medic, decorated her fire helmet with glue-on pearls, flowers drawn with colored Magic Markers, and the balloon-shaped letters "EMT" in gold glitter. It looked like a schoolgirl's notebook, but she figured the decorations might calm some of the injured people she helped. Jen hadn't had time to decorate her own helmet. These are stout women — strong enough to handle livestock or help carry accident victims, and good-natured about the hard work that comes with life on a farm and a volunteer fire department. Neither had ever fought a wildfire.

"I'm glad to see you here!" Jen shouted when she walked into the firehouse a few minutes after her sister-in-law had arrived. The women grabbed their flame-retardant coats, pants, boots, and helmets and pulled out in the department's ancient gray Suburban as fires erupted all over the Colorado plains.

Damon and Darin doused a farmer's burn of some felled trees that the wind was blowing out of control, but soon everyone who was available was called to the first fire that came across their radios. The blaze was named for the town that had once existed 60 miles south of Wages, where the wind had blown down a power line that had then snapped sparks into the dry grass — Heartstrong.

A MAP ON THE WALL of the Yuma County Sheriff's Office described Heartstrong as also having the names Headstrong and Happyville. Cleve Mason was living up to the first of those alternate names in 1920 when he opened a post office there — without permission from the postal service. He used the second name for his office. The next year Mason's stubbornness paid off when the postal service approved the Happyville Post Office, which distributed mail until 1940. Today there's no sign of a town with any of those three names. "There's nothing there," Sheriff Chad Day said. "Nothing."

Wintertime wildfires threatening invisible towns were just the latest surreal development in Day's service as Yuma County sheriff. "Nothing about me in this job makes sense," he said.

At 34, Day was the third-youngest sheriff in Colorado history. He grew up on his family's farm in the county and studied agriculture at Colorado State University, in Fort Collins. After graduation he worked at the Larimer County jail, got promoted to deputy, and won a hard-fought campaign against the sitting sheriff. He moved his wife and kids back to his family's farmstead near Wray, taking a $20,000 pay cut and a big increase in responsibilities. But things started to fall into place. His wife, who had once resisted even visiting the northeast corner of Colorado, embraced the move. They found a palatial house and made an offer that they considered insulting but was what they could afford. The sellers accepted, citing the young sheriff's reputation and his family's long legacy in the county as the reasons. Sheriff Day's new home had a full-sized racquetball court in the basement.

Short and fit, with a close-cropped beard and hair, the sheriff still looks like a college athlete. He wears his sheriff's office coat like a letterman's jacket. He'd been in office just over a year when, while he was coming home from church with his family, the dispatcher called in the Heartstrong Fire.

He headed for the fire, then saw a wall of smoke rising in the distance and turned back to set up a command post next door to the Yuma Police Department. By then his radio had announced the first farmhouse burning down.

Duaine and Lucie Eastin's farm was at County Roads 26 and M. The fire came on so fast that when their son and daughter-in-law came to help them prepare for it, they could do little but pack a few things and "get mom and dad out of here."[5]

Less than an hour later the blaze rolled over George and Florence Pletcher's farmstead, three miles north of the Eastins' at County Road 29, bringing down the home, barns, and sheds like a wave washing away sand castles on the beach.

Elmer Smith had been chief of the Yuma Volunteer Fire Department since the first of the year and was one of its 34 firefighters before that. Smith drove a truckload of water and air packs to firefighters trying to protect the Eastins' home. The fire was already leaping out of the eaves and curling the tin roof.

"We gotta get these guys out of here," he announced.

After convincing Eastin and the six firefighters trying to save the home he had lived in for 75 years to retreat, Smith drove out in front of the blaze to figure out how to fight it. "I had no idea where this fire was," he said.

It was like trying to drive through a black and gray blizzard. He could hardly see through the cloud of sand, dirt, smoke, burnt grass, and cornstalks flying in the relentless wind. On a road that should have held back the fire, the flames just leapt over Smith's car to ignite the field on the other side.

Smith found himself coordinating 84 fire trucks with volunteers from different departments. Others from Nebraska and Kansas backed them up. The fire burned to within two miles of the town of Eckley, and deputies and state troopers coordinated the evacuation of the town from the Silver Spur Bar and Grill.

Sheriff Day called for a mandatory evacuation over about 40 square miles, but it seemed that most people were headed the wrong way. Deputies and firefighters couldn't get to homes that needed protecting fast enough because the roads were so choked with people and cars. Some left their vehicles on the road and ran into fields to help farmers round up livestock, but most were rubbernecking at the oncoming fire. Drivers blinded by the blowing sand and smoke, or distracted by the spectacle, crashed into ditches. One car rear-ended a fire truck. As the blaze overtook the Pletcher farm, their cattle joined the cars filling the road.

"WE'RE NEVER GOING TO GET ANYTHING DONE but driving today," Jen Struckmeyer said in the passenger seat of the Suburban as her sister-in-law navigated the chaos. "Let's get this thing done!"

To the east rose a wall of smoke more than a dozen miles wide and hundreds of feet tall. Flames spread over some 38 square miles.

Windblown sand and embers left pits in windshields and blasted a light off the side of Deputy Curtis Witte's patrol car.[6] One rookie Yuma firefighter said he only realized how hot it was when a chicken fell over dead at his feet. A state trooper reported seeing cattle with their hides burning and their skin falling off. He wanted to stop and shoot them to end their suffering, but there were still homes to evacuate. By the end of the day about 50 head of cattle had perished in the fire, and nearly 200 more were missing.

Farmers turned on their irrigation sprinklers to hold back the blaze, and cattle huddled in the shelter of the mist. Others drove tractors pulling disc harrows miles ahead of the approaching fire, trying to contain the blaze with a line of dirt. Grass fires on the plains are so fast the armies of firefighters that battle forest fires can't respond quickly enough.

Knowing that farmers were violating his call for an evacuation and risking their own lives, Sheriff Day was torn. "The only way to get them out of there is to put them in handcuffs and drag them out," he said. "I'm not going to do that."

Rows of plowed soil, however, couldn't hold back the flood of fire.

THE GRASSES ON THE PLAINS are known as one-hour fuels to firefighters — so fine that heat, lack of moisture, and wind can dry them to the point of combustion in 60 minutes. Bunchgrasses in Colorado's sand hills grow in clumps, and their sparse spacing usually slows fires. But months of drought, days of temperatures a dozen degrees above normal, and hours of relentless wind had primed the grass, corn, and wheat stubble to burn, and too little snow had fallen over the winter to mat them down, leaving upright stalks that ignited like matchsticks. Fast grass fires usually just torch the tops of corn and grass stalks, but during the Heartstrong Fire these stalks burned like fuses into the ground. In some fields it would take years for soil moisture to return to normal levels.

With the wind behind them, flames stretched out horizontally to ignite grass and sagebrush. Yucca plants, which often resist fire, burned like candles. Gusts of wind bounced fiery tumbleweeds like flaming tires.

In the hour after it started, the Heartstrong Fire raced six miles up the plains.

· · ·

"THIS FIRE'S PROBABLY 10 TIMES WORSE than I've ever seen," Damon Struckmeyer thought as he approached it.

The three trucks driven by Struckmeyers — Damon and Darin in the fast-attack fire truck, Pam and Jen in the Suburban, and Del driving down from Sterling — tried to stay in touch, but there was so much radio traffic, they had trouble getting through to one another.

Pam and Jen were waiting in a makeshift staging area with some other trucks when Damon and Darin pulled in. Damon guessed that leadership on the fire was still coming together, so he didn't contact incident command for instructions.

"Flames — let's go after 'em," Damon said. "There's a need right here, so let's just jump in and go."

Pam climbed into the fire truck's passenger seat. Darin and Jen strapped in on the platform between the cab and the water tank.

The wind was so hot and fast, it could have been coming from a hair dryer. Jen and Darin hunched behind the cab to shelter from it, and squeezed their eyes tight against the smoke and debris blowing into their faces.

Where the flames came up to the road, Damon angled the truck so Jen could spray them, but half the water just blew back onto her. When she couldn't bear the heat, she would pound on the roof and shout for Damon to back off. "This is absolutely horrible," she said.

Damon cut a barbed wire fence to chase the fire into a field. He was just a few hundred feet from a gate, but it was so submerged in smoke he couldn't see it.

DEL STRUCKMEYER ARRIVED at the staging area in time to see the fast-attack truck turn into the field and vanish in the smoke. He cursed himself. On his drive down he'd helped a stalled water truck. If he hadn't delayed, he would be riding with his wife. He did not like the idea of her heading into the fire without him.

DAMON DROVE ALMOST IMMEDIATELY onto a series of dunes hidden in the smoke. The sugar sand was so dry it was like trying to drive in freshly milled flour. When the wheels lost traction climbing, Damon backed up and tried another slope, hoping to get behind the flames. From the top of one hill, he'd plan how to get over the next, until suddenly it seemed like they had arrived at the center of creation. Fire whirls — tornadoes of flame — spun into the sky with the regularity of telephone poles.

"Those flames are taller than the power lines," Damon thought. "It's just a constant line of them. I didn't realize that it was moving like this."

The fire was about 200 feet away when Damon decided to retreat. He headed northwest toward the road they had come in on, with the fire closing in.

On his way up a hill, the truck spun out in the sand. As Damon backed down, he heard a *whumpf.* The truck sank up to its axles. Damon shifted between drive and reverse, but the truck sank deeper.

Despite the debris that blew in his eyes when he stepped from the cab, Damon could see the flames as bright as the sun rising behind him. Before he could close his door he felt the wind shift to the northwest. The fire turned toward him, as if it had just seen its prey. He shouted a warning to Darin and Jen, who were unsheltered on top of the truck and could hardly hear over the jet-engine roar of the fire.

Jen jumped from her perch, but her right boot got caught and came off. She landed on the sand with her foot protected only by a sock. The flames were 50 feet away and coming on fast.

"We're all done," she thought. "This is hot and this is fast, and there's no way of getting out."

She ran past the front of the truck, slamming the driver's door shut as she went by.

"Maybe that will keep the fire from getting to Pam inside the cab," she thought.

Running in the sand was hard, particularly with one boot.

"If I . . . just get to the fence, then I'll be okay," she thought "I can get at least into the ditch and get away from it."

Then the blast from the fire hit her from behind. She crashed face-first into the ground.

"I'm going to die," she thought.

WHEN DAMON HEARD THE DRIVER'S door slam, he guessed Jen had climbed into the cab. He ran to the compartment where he kept his fire coat and tried to pull it on, but the wind was blowing so hard, he couldn't get his arms into the sleeves, so he dropped it and climbed onto the platform with his nephew. He grabbed a hose and set it to spray a wide fog and blasted as much water as he could into the air over their heads.

The flames were on top of the truck less than a minute after the sand had swallowed its tires.

Pam tried to get out of the cab to bring her sister-in-law inside it, but the wind pressed her door closed like rushing water. She saw Jen fall with her bootless foot in the air. Then the smoke blocked the view. She looked out the back window as her husband's face vanished into a sphere of fire and smoke that ballooned above the truck, sucking the air from the cab.

"We're going to be just spots in the sky," she thought.

Without his coat, Damon could feel the flames burning his elbow, his shoulder, his face. Cinders blew into Darin's eyes, all but blinding him.

The flames splashed over them in big waves, then rolled away.

"A bull ride . . . lasts about eight seconds," Damon said. "This was longer."

The heat reached Jen, facedown in the field, before the flames. She put her booted foot over the unprotected one, but her sock melted onto her skin. The flames chewed through the hip of her fire-resistant pants, then seared the skin they exposed. It burned her arm through the elbow of her jacket. The visor on her fire helmet turned to goo. The lenses of her glasses crinkled and yellowed. She could feel the skin on her face cooking.

"Lord, if you're going to take me, take me now," she prayed. "But if not, you've got to stop this fire. You've got to stop the burning."

Jen took a breath as the flames surrounded her. Superheated gas reached down her throat to take it back.

AS THE HEARTSTRONG FIRE MARCHED away from the Struck-meyers, it left behind a fog of smoke and soil blowing over the charred ground like a battlefield pummeled by artillery. Melted plastic dripped from the fire truck. The sand that had swallowed the vehicle's tires, trapping the family in the fire, may also have helped them survive it.

With the truck sunk up to its axles, the flames couldn't get underneath it to ignite the engine or fuel tank.

Damon jumped from the platform and pulled open the driver's door. "You girls okay in here?" he asked.

Pam said that Jen was out in front of the truck, but even with his wife's guidance, Damon couldn't find his sister-in-law until the wind lifted the veil of smoke long enough for him to see her lying facedown on the ground. She wasn't moving. Damon and Darin ran to her and could see the holes burned through her fire pants and jacket. The foot exposed when she lost her boot was charred. Her hands had vanished.

But she wasn't dead.

Damon hosed a mist over her to cool her. She had pulled her hands into the sleeves of her coat to protect them, but some of her fingers were burnt to the bone. They felt like balloons about to pop, and she begged her family to ease her pain. Pam poured a bottle of water down Jen's sleeves to cool her hands, but the fabric was still as hot as a griddle, and steam hissed through the coat, poaching the skin on Jen's arms.

Damon hooked her up to an oxygen tank, then used the backboard to shelter her from the blasting sand. Pam radioed for help while Darin turned on every light that worked on the truck to make them easier to find.

WHEN DEL STRUCKMEYER HEARD that the fire had burned over his family, he was just half a mile away from them. He started searching the fields, but his truck bogged down in the sand. Wages firefighters driving the department's six-by-six picked Del up and continued the search. They found truck tracks near the fence but struggled to follow them in the wind-scoured sand. By then, from listening to Pam's and Damon's desperate calls on the radio, he'd figured out it was his wife who was most seriously burnt.

Half a dozen trucks from other departments joined them, but in the thick smoke it was like searching in the dark.

Some 40 minutes after Pam Struckmeyer's call for help, a gust of wind punched a hole in the blackness, and a truck of Yuma firefighters caught a glimpse of green paint. They had been within 50 yards of the Struckmeyers for half an hour without seeing even the lights flashing on the truck.

Although ambulances were waiting just a mile away, heavy fire blocked the road, forcing the Yuma crew to drive 20 miles to get to them. It took the better part of an hour to get the Struckmeyers to the Yuma District Hospital, but they spent only a few minutes there before doctors sent them to the burn unit in Greeley, again by ambulance. It was a two-hour drive, but the winds were too strong for helicopters to fly.

By the time they arrived in the burn unit, the Heartstrong Fire was dying. The winds driving the blaze calmed at around 8:30 that night. Sand and soil held aloft for hours fell to earth in deep drifts, smothering the flames. Green crops in irrigated winter wheat fields wouldn't burn when the flames came on them.

"We saved what we could," Chief Smith said afterward. "But I really think Mother Nature put that fire out."

And while the flames might be out, he said, they weren't gone.

"We're going to see this again."

IN GREELEY NURSES RACED JEN to a shower and laid her on a table beneath yet another set of nozzles. "I've got these blisters the size of balloons on me," she thought. "What are they going to do to me?"

When they started scrubbing, the first of the daily showers she would receive for months, she screamed. "If this is what I'm in for, take me back to the fire!"

Doctors found third-degree burns on her foot, fingers, and hip, and second-degree burns over 25 percent of her body. Her legs would require skin grafts, and her fingers would require surgery to repair the tendons. Most of the five toes on her right foot would be amputated.

After a few days the superheated smoke she'd inhaled took its toll on her lungs. She suddenly couldn't breathe without a respirator. Her condition turned critical. Doctors stabilized her, but she would spend months in the hospital and a rehabilitation center. If she had been older or less fit, they told her, she probably wouldn't have survived.

Despite the Struckmeyers' suffering, less than a week after it ignited, the Heartstrong Fire seemed like little more than a fuse to the most explosive series of fires in Colorado history and the West's new endless season of wildfires.

5

RED BUFFALO, BLACK DRAGON

A QUARTER CENTURY BEFORE THE HEARTSTRONG FIRE, I stopped my car on the shoulder of a gravel road in the middle of Kansas, in the middle of the night. From a distance the "red buffalo" glowed like a miniature sun rising behind the low hills. Occasionally a wave of smoke backlit by the ruddy glow splashed over a hummock like dust on a bison's back in front of the setting sun.

When it came into closer view, it was shaped more like a massive snake than a buffalo, or perhaps magma seeping from a mile-long crack in the dark plains.

Natural fires often burn along prairie hilltops, where lightning strikes are more frequent and the grass dries out earlier in the season than in the moister draws below. Humans, however, usually set their fires lower, where the grass grows thick and matted, to improve the forage and enrich the soil.

Standing in the Flint Hills of Kansas, I was in the heart of the largest spread of the Great Plains to have never been plowed. Some 20 miles to my west, outside El Dorado, is the site where, in 1872, a grass fire overtook the Wheaton family, killing a young mother and her son.[1] Her husband, trying to save them, endured burns so severe that his hand had to be amputated, while his daughter survived unscathed in the family's oxcart. A few miles to the north was what 10 years after my visit would become the Tallgrass Prairie National Preserve, as twenty-first-century residents saw the grassy plains less as a barren wasteland filled with peril and more as a glorious park. By then less than 4 percent of the original 170 million acres of prairie grasslands that spread from Indiana to

Colorado between the Mexican and Canadian borders remained as they were before the western expansion of the United States.[2]

From where I parked, the "sea of grass," growing from what was millennia ago a sea of water, rolled away in long, low waves into the darkness. The flames rose two to three feet high and lazily worked their way east. With a puff of breeze the grass rustled, and flames kicked into the air and shuffled a few feet farther across the plains, like a chorus line stepping forward on a stage. The fire was so unthreatening that a few cattle calmly munched grass nearby, slowly ambling away from the flames when they danced too close.

Native Americans stoked roiling grass fires to drive buffalo toward their waiting spears or over cliffs,[3] but the bison often seemed as unfazed by small burns as the cattle I saw standing by the flames in Kansas.

At Wind Cave National Park, at the south end of the Black Hills of South Dakota, Carl Bock, now a professor emeritus at the University of Colorado, saw a group of buffalo caught by a fire. National Park Service employees had lit the fire not to herd the bison to slaughter, but to burn away some of the pines and improve the animals' forage. They had seen, by comparing photographs taken during one of Lieutenant Colonel George Custer's expeditions through the region with modern photos taken from the same vantage points, that ponderosa pines were overtaking the mixed-grass prairie of big bluestem, little bluestem, and side oats grama. Bock recalled that the Park Service had lit that fire on the day John Hinckley Jr. shot President Ronald Reagan in Washington, D.C.

"The bison seemed to be drawn from the park to the burn," he told me. "As soon as it passed, they would begin to lick the ash. Just by chance a group of bison were trapped on a knoll with fire coming at them from 360 degrees."

The buffalo were "really relaxed," waiting until one cow saw a safe route and charged through the fire as though she was running across a river. The rest followed her, single file, and on the other side of the ribbon of flame stopped and lowered their heads to lick the just-burnt ground while it was still hot. Bock wondered whether the thick hair that bison carry on their heads, shoulders, and chests, which they do not shed, is an adaptation for confronting fire on the plains. Only one of the bison that Bock saw run through the fire was burned — a young bull that broke its neck rear-ending the animal it was following. Its backside was burned to the bone, but its front had only singed hair.

At the end of the last ice age, much of the Midwest was covered by forests. Between 8,000 and 9,000 years ago, the warming that pushed back the ice also turned forests, which are less tolerant of heat and drought than grasses, into prairies.

As opposed to the prairies farther west, where the rain shadow of the Rocky Mountains keeps the shortgrass prairies just ankle high, in the tallgrass prairies of eastern Kansas big bluestem can grow 10 feet tall and sink its roots 12 feet into the earth. Little bluestem, switchgrass, and Indian grass don't grow quite so high, but are just as nutritious for grazers. As rich as the grasses are, three-quarters of the life on the prairie is underground, sheltered from fire. No other U.S. ecosystem sucks as much carbon from the atmosphere as the "lungs of the nation," the prairie grasslands. And only Brazil's rain forests, the "lungs of the earth," are a more complex or diverse biome. Among the tallgrasses are more than 300 species of flowers and forbs — evening primrose and wine-cup, blazing star and ground plum.[4]

Foxes, bobcats, weasels, minks, pronghorns, and even badgers scraped out a living in the Flint Hills. For bison, the tallgrass prairie was about the best meal they could hope for. They achieved their greatest densities crowding into it to munch on new grass growing from recently burnt land in the spring and summer.[5] By autumn some stalks were so tall, the herds seemed to swim through them. Though the bison once numbered more than 28 million, the last wild bison in central Kansas had been gone for a century when I stopped to stare at the flaming grass. The 13 bison the National Park Service reintroduced to the tallgrass prairie decades after my visit would never have the impact on the ecosystem that the great herds did. After the 1880s, the demise of the bison allowed grasses, and the fires they fed, to flourish on the plains.

For a time whites attempted to eradicate the red buffalo from the plains in much the same way they had the brown one.

"Prairie fires are the crying evils — the curses — of our state, and there must be protection against them," the *Emporia (KS) News* charged in the 1860s. "We consider that the evil effects of prairie fires . . . are greater than all the drouths, hailstorms, and hurricanes that ever have or will visit us."[6]

Most plains communities in the late 1800s viewed natural grass fires with the same dread as they did tornadoes, and many outlawed man-made fires on the plains. But in the Flint Hills, where the rocky soil made farming difficult, the nutritious grasses were a cattleman's dream. The forage required fire to be salubrious. Ranchers soaked fist-sized balls of burlap in kerosene or diesel fuel and dragged the flaming orbs behind their horses and trucks. Winter and early-spring burns raise the pH level of the soil.[7] Late-spring fires lower it. After a fire, sunlight on the blackened ground encourages microbes that recycle nutrients into the soil. Steers dining on the fresh growth that follows a fire gain 10 to 12 percent more weight than those eating old stalks. For more than a century ranchers from as far away as Texas, Florida, and even Mexico have sent their cattle to the Kansas and Oklahoma hills to fatten up on fire-nourished grasses.

Before the vilification of prairie fires, ranchers burning away the old and excess grass that the exiled bison had once mowed down were not the only settlers domesticating the red buffalo. Immigrants to the plains burned to make it easier for them to hunt and to create habitat in which their prey could proliferate. They burned to make space for houses. They burned to deprive pests of homes.

Until about 150 years ago the pests in question were often human. Native Americans and European settlers set fires to drive one another away, to starve their pursuers' horses of grass, to push enemies into ambushes, to destroy crops, and to make land uninhabitable.

"The Indians now set fire to the prairies and woods all around us," wrote Charles Augustus Murray in his 1839 book, *Travels in North America*. "These malicious neighbors were determined to drive us from the district; they evidently watched our every motion, and whenever we entered a wood or grove to hunt, they were sure to set the dry grass on fire."[8]

The red buffalo was the most useful piece of the wild ever lassoed by the settlers of the plains. But at its heart, it is still wild, and it often slips its lariat, jumps a fence, or just overpowers the hand that ignited it. In 2010 Celia Harris, a retired schoolteacher, was helping out on a controlled burn in the Flint Hills near her home. The red buffalo killed her the way cattle can crush a rancher — trapping her against a fencerow.[9]

Downed power lines, sparks from machinery, cigarettes tossed out car windows, and dry lightning still ignite charging herds of flame. The vast, thundering stampedes of the red buffalo gallop across the plains beneath roiling clouds of smoke, dust, dirt, and cinders. While our national consciousness holds the image of a wildfire as a flaming mountain forest, the blazes of the plains are faster and, on occasion, more devastating and deadly.

In March 2006 the fires known as the East Amarillo Complex burned nearly a million acres of the Texas Panhandle in four days. Grasses had proliferated during an unusually wet year, but then dried to kindling when the Panhandle fell into "extremely critical drought" the next year. Two blazes ignited within minutes of each other when high winds blew down power lines more than 20 miles apart. The fires merged to race across 45 miles of prairie in nine hours.

The Texas blaze overran four drilling company workers when their truck was trapped in a ravine as they tried to outrun the fire. They were among 12 people killed on the first day of what was the largest and deadliest wildfire in the United States that year. Smoke smothered Interstate 40, leading to a nine-car pileup that killed four people from Oklahoma. In addition to the human casualties, 4,296 cattle and horses perished. Ranchers reported flames so fast, they incinerated steers where they stood. More than 2,000 miles of fencing and 1,000 utility poles burned.[10]

Three years later, in Oklahoma, J. C. Myers, a volunteer with the Union Chapel Fire Department, died when he raced his fire truck onto a smoke-filled road and crashed head-on into a pickup driven by his son, Juston Myers, who was responding to the same fire.[11] Fires plagued Texas throughout 2011, culminating in a dozen firestorms that broke out across the state on Labor Day. Just one — in Bastrop, near Austin — destroyed nearly 1,700 homes.

No matter how wild it runs, however, the red buffalo is barely a match smoke compared to the Black Dragon, the fire that exploded in China and the Soviet Union during the same spring when I stopped to look at the lazy flames in the Kansas grass. The two fires seemed like a drive-in double feature. The red buffalo was a Western. The Black Dragon unfolded like a monster movie.

The world's largest coniferous forest, the Da Hinggan — "Black Dragon" in English — fills the northeast corner of China like a flexed arm reaching into Siberia. Its larch, spruce, and pine sprawl from North Korea to Mongolia. Before the 1980s the vast wilderness had just a handful of roads and settlements, but China's exploding growth fueled a hunger for lumber that led the nation to cut a railway through the heart of the forest that held half its timber reserves.

The tree harvest was overseen by the Black Dragon "family," a "Mafia-like corporation that ran the forest as a kind of pleasant feather bed with little regard for safety and knowledge of firefighting," wrote Harrison Salisbury in his book *The New Emperors: China in the Era of Mao and Deng.* They built comfortable homes but let fire towers fall into disrepair, and left behind towering piles of shattered wood and brush after their harvests.[12]

In 1987, when a worker refueling a brush cutter slopped gasoline onto the forest floor, he unleashed a horror like no living human had ever seen. Within hours flames hundreds of feet tall leapt through the forest. The Black Dragon Fire burned more than 18 million acres in Manchuria and Siberia — an area nearly the size of New England and 10 times what the infamous Yellowstone fires would burn in the United States the next year.[13]

On the Chinese side of the Amur River at least 60,000 people, including two armies of soldiers, battled the blaze. Battalions of artillery shelled the sky with dry ice in an effort to compel the clouds to rain. On the ground tens of thousands of Chinese workers tried to bat down the flames with little more than flyswatters. Despite the explosions in the sky and the flagellation of the earth, the Black Dragon devoured 10 percent of the world's coniferous forests.

The fire melted telegraph and telephone lines and then washed over villages that, cut off from communications, didn't realize the size and ferocity of the beast in the sea of smoke. The conflagration killed at least 220 people, an astonishingly low number considering more than 33,000 people lost their homes.

The Black Dragon's breath would be felt throughout China and around the world for decades. The fire destroyed nearly $6 billion in timber, driving an explosion of lumber imports into China that sped deforestation in a dozen other countries. A year after the fire, the worst dust storms and sandstorms in decades plagued Beijing, more than 1,000 miles from the charred forest.

"When Beijinger met Beijinger there was one word on their lips: Heilongjiang! The Black Dragon fire!" Salisbury, then an octogenarian reporter for the *New York Times,* wrote. "The fire had begun to take its climatological toll."[14]

The fire, Salisbury wrote, "possess[ed] an almost mystical quality, representing a symbol, which I did not entirely understand, of the direction of our future."

Yet even he couldn't predict that in the following 25 years, sequels to the Black Dragon horror show would burn on every forested continent.

ON THE OTHER SIDE OF THE PLANET from where the Black Dragon had raged, I leaned my hip against the fender of my car and watched the red buffalo slowly graze on the Kansas plains. I'd stopped to watch prairie fires before and would again. Before I quit smoking, I'd light a cigarette as I watched the blaze. I didn't recognize the connection between the ember held by my lips and the one holding my gaze. Yet the ability to apply smoke to my lungs and also to return the grass's nutrients to the soil are just two of the myriad benefits that fire bestows on *Homo sapiens.*

Fire forged and molded the vehicle that I was leaning against, and created the combustion in the engine that brought me to watch the prairie burn. Fire allows us to light the night, drive back the cold, and keep predators at bay — distinct advantages over every other animal. For many anthropologists, it was our alliance with fire that made us human.

Fire created the big brain that allowed me to contemplate the scene before me. Research by Rachel Carmody at Harvard has shown that cooking meats and vegetables releases nutrients, starches in particular, that aren't digestible in the foods' raw state. Cooking makes meat easier to chew, swallow, and digest, and gave our ancestors access to foods densely packed with nutrients and animal proteins that allowed our brains to grow some 86 billion neurons — more than twice a gorilla brain's 33 billion and three times the 25 billion in the average chimpanzee's.[15] Our brain's size advantage over our closest primate relatives', however, requires a lot more fuel. The human brain uses 20 percent of the body's energy when it's resting — double the portion that other primates' brains burn. To satisfy our caloric needs with uncooked foods, we would have to eat nine hours a day.

Richard Wrangham, an anthropologist who works with Carmody at Harvard, argues that one way cooked food helped our brains to get big was by allowing our guts to get small. We required less heavy-duty intestines to process food already broken down by cooking, and our bodies diverted the energy no longer required by our more efficient, downsized guts to our rapidly expanding brains.

If I run my tongue over my teeth, I can feel the impact of fire in my mouth. Humans' "dental chaos" — our teeth are unusually disorderly compared with other animals' — is likely the result of our jaws and teeth shrinking at different rates in response to the need to chew cooked food less, according to Peter Lucas at George Washington University. Chimpanzees spend about half their days chewing, while humans masticate for less than an hour each day.[16]

Wrangham believes that the time saved from not having to chew raw vegetables freed humans to hunt more — another way in which cooking gave us an evolutionary advantage by adding more meat to our diets. He theorizes that our ancestors began using fire up to 1.8 million years ago. That's almost a million years before the oldest known evidence of fire use by humans, which was found in South Africa's Wonderwerk Cave. The anthropologist points to the evolution of the greater honeyguide, an African bird that leads humans to honey with specialized dances and calls. Humans follow the birds to the hive, then use smoke to disable the bees and gather the honey. The birds also get a payoff in the hive's wax, which they find delicious. Research by Claire Spottiswoode at the University of Cambridge indicates that the honeyguide required more than a million years after humans began using fire to evolve the guiding behavior.

Plains Indians drove bison with fire, but also noticed the animals preferred grasses growing on prairie that had recently burned. So the hunters started fires not only to herd the bison but also to keep them close with the improved forage that followed the blazes.[17] Early Cherokees burned grasslands near their villages so they didn't have to travel as far to hunt the animal that provided their food, clothes, and tools.

There have always been dragons in the mountain forests, raging huge and red on the jagged horizon, and they have always fed humans' imagination. But it is the red buffalo that has fed our stomachs, lit and heated our caves and camps, formed bones and wood into tools, melted ores into metals, heated water into steam to spin electricity from our turbines, and fired up the engines in our cars and planes. Fire in its myriad forms remains humanity's most powerful tool and technology.

But in the eons that we've been herding, harnessing, breeding, and butchering fire, we have also marked every inch of the earth with our soot, flames, and hunt for fuel. Black carbon from human fires darkens ice in the treeless barrens of the Arctic and Antarctic, speeding the melting of glaciers. Deserts are filled with forests we've planted for timber or fuel. Carbon dioxide from our cars and power plants is changing the chemistry of the atmosphere and the ocean, heating the planet as an unintended consequence of our fires. And in woodlands around the world we've attempted to eliminate fire as if it were a wild animal that we could exterminate. But now the dragons have returned — larger, hungrier, and, it seems, at least metaphorically, enraged by our efforts to harness them.

As I drove away from the grass fire in the Flint Hills, I saw a pair of headlights on the prairie beaming into the line of flames, while half a dozen taillights around it glowed like cigarettes. Perhaps the drivers were just admiring the same spectacle I was. Or they might have been the ranchers who had started the blaze. In either case, they were keepers of a flame starting to strain against its shackles.

And, I would learn, there are other shackles associated with wildfire.

6

CRAZY WOMAN

I DIDN'T INTEND TO GO TO PRISON. Certainly not to get overrun at a fire.

In 1986 I had just taken a job as a newspaper photographer in Connecticut when I heard about a grass fire over the police scanner. It was the first wildfire I ever responded to, either as a journalist or a firefighter. It couldn't be anything like the flaming mountains of the West I'd seen photos of, or the fires I'd seen ranchers and farmers set on the Kansas plains, but I was nonetheless eager to get in on the action. I drove to a road from which I could see the smoke plume, then climbed over two barbed wire fences, with my cameras swinging off my shoulders.

When I could see flames in the grass and men digging a line around it, I jogged close enough to start making pictures of them. I'd hardly had a chance to snap a frame before I heard yelling. When I looked, I saw that the shouting wasn't coming from the group surrounding the fire, but from a man in a uniform who was marching across the field toward me. As he came closer, I could see a badge.

"What the hell are you doing in here?" he screamed. "You've got to leave. Now!"

He was a guard at the nearby minimum security prison—built like a truck. The fire was burning on the grounds of one of Connecticut's correctional institutions. The firefighters were inmates. The press was rarely welcome there, and never without an appointment.

I stuttered an explanation but didn't manage to get a whole sentence out. He came at me like a football player, grabbing me by my armpits and dragging me away as if I were a toddler. I realized he was handling me the way he was trained to restrain the inmates.

"This is a fucking prison!" he shouted. "You can't just walk in, and they can't just walk out."

As his anger exploded at me, the blaze behind him blew up. A ball of flame started rolling across the waist-high grass toward one of the firefighters. It engulfed wooden fence posts and reared above him. He ran toward me to escape it. I raised my camera and tried to get a shot off past the guard's legs, but my rough-and-tumble removal made framing and focusing impossible.

The guard noticed my telephoto lens banging against his legs and looked down to see me trying to take a photograph. Then, looking over his shoulder, he saw the prisoner and the flames. I expected him to grab my cameras, or at least block my lens with his hand, but he lifted me to my feet and pushed me to face the man running for his life. The guard stepped back as I pressed my finger down on the motor drive. The camera captured a sequence of black-and-white photos of the prisoner running, still carrying his shovel, with the fire exploding behind him. The blaze rolled faster than the inmate could run, growing larger and faster as it devoured more of the thick, high grass. Just as it seemed it would swallow him, the wind subsided and the flames fell and retreated, like a wave crashing on a beach and then falling back into the ocean. Once the prisoner had escaped the flames, the guard grabbed me again, resumed his curses, and pushed me back to the fence where I had come in. I never got a chance to thank him for man-handling me into the position from which I took the photos. I never learned his or the inmate's name.

But I did learn that inmates fight wildfires. And most of them are fighting blazes outside prison fences.

When I dug deeper into the use of inmates to fight wildfires, I learned the California state prison system established road camps in 1915 to put low-risk inmates to work. When World War II drew away most of the state's forestry workers, inmate firefighters battled blazes from 41 temporary camps. A century later, nearly 4,000 minimum security inmates — men, women, and juveniles — were serving on 200 fire crews in California. With the number of California wildfires increasing, as many as 37 percent of the inmate firefighters working in California in 2015 had been violent offenders, according to California corrections officials.[1] Most western states have similar programs, putting prisoners in fire crews instead of chain gangs.

"Maybe," I thought when I learned of the programs, "if they're so desperate for firefighters they're putting prisoners on the fire line, there might be an opportunity for me."

YEARS LATER, IN 2003, a lightning strike on Wyoming's Crazy Woman Mountain was my ticket to the front lines of the nation's war on wildfire. After five years of drought, it was virtually guaranteed to start a fire. The bolt blasted the scruffy forest of ponderosa and piñon pine and juniper, and the fire spread quickly through the heaps of bone-dry tinder on the steep slopes.

In New England, western wildfires seemed remote, but burning federal land presented opportunities for firefighters across the country. The winter before the fire, I'd signed on for a week's worth of training and joined a crew organized by the Connecticut Department of Environmental Protection.

By early July 2003, 28,000 people — all the civilian resources available — had been called into the battle by the federal government. They were joined by U.S. Army troops and foreign firefighters from as far away as New Zealand. That month I bounced from fire camp to fire camp over hundreds of miles of Rocky Mountain roads in a school bus. I knew that my weeks working on the fire line would educate me, but I didn't anticipate that schooling would be as much in economics and politics as it was in fuels and fire weather.

I arrived to fight my first forest fire with a crew that subscribed to the message of Smokey Bear, which had been around since before any of us were born: wildfires are a scourge that will destroy our forests if we don't stop them. I left believing that most forests in the West were sick and overcrowded because we'd put out the fires that functioned as their immune systems and gardening crews. In both cases I was subscribing more to mythology than to science.

"What's this?" our normally quiet crew boss shouted on the first fire line I would cut in the Rockies.

A few skeletal trees burned behind him, but most of the fire scene was hidden behind clouds of white smoke.

"The safety zone?" one of his huddled crew questioned.

"That's right," he yelled. "The black is the safety zone."

In wildfires people live by embracing death. The blackened earth — land that the fire has already burned — is often the only place where a person can escape the flames.

Like all wildland firefighters, we learned 10 Standard Firefighting Orders and 18 Watchout Situations in the classroom, but remembering them when a fire goes bad is like trying to recite the Bill of Rights in a hurricane.

And the list of perils is far longer than those.

Snags — trees weakened by fire or disease — can fall at any time. "They'll drive you into the ground like a tent stake," one of the veterans on my crew warned me.

Junipers split and twist when they fall, launching shards of wood and branches in all directions. Stumps burn into the ground, leaving ash-covered pits of embers that can swallow a firefighter's foot like burning quicksand. Hot ash runs like water, sometimes delaminating boots and blistering feet.

A slight drop in relative humidity, a jump in temperature, or an increase in wind speed can rouse smoldering embers into a racing wall of flame that can easily overrun a firefighter standing in the wrong place. Smartphones, satellites, handheld electronic weather meters, and portable weather stations give firefighters the most technologically advanced weather prediction capability. Still, our squad bosses "spun weather" — measuring relative humidity with glass bulbs hanging from chains they swung through the air. They tested wind speed with anemometers they held above their heads, and monitored thermometers hanging from their packs. Their wariness of tech was akin to the skepticism many had toward the federal bureaucracy. It often proved to be well-placed.

Just as we finished our safety briefing on that first fire line, a slurry bomber unexpectedly rose into view and dropped its load of retardant before the crew could get out of the way. The red mud splashed over us, stinging our skin and staining our clothes. A more direct hit would have knocked us down, or worse, but most of the firefighters relished the red blotches on their clothes and the thrilling sight of the bomber.

"You got the red badge of courage," one told me as he slapped my back. "Some guys come out three or four times before they get that."

Our three squads lined up like centipedes that crawled slowly along the edge of the burning land to "hit and lick" a line around the blaze. With every step, each man chopped the vegetated ground with an axe, hoe, or rake, then moved up the line. By the time 20 firefighters had passed, there was an 18-inch ribbon of mineral soil between the black of the fire and the green of the forest.

Water is a luxury when fighting wildfires. Dirt is the suppressant of choice. We shoveled it onto flames to smother them and stirred it into embers to cool them.

Sweat and smoke stung my eyes, and after a few hours my back and shoulders ached. But swinging an axe into wood and dirt, and walking back along the fire line we'd cut through the woods, left a tangible sense of athletic accomplishment. I considered the disappointment so many crews feel when a fire burns over a line they spent days cutting, and their backbreaking work vanishes in a charred forest. But for the moment, as I stood with my crew between the fire and the forest, I felt proud, like I was a piece of something both particular to that small mountaintop and part of a massive, nationwide effort.

I learned more about fire in my first hour on the line than in all my training, although even my weeks on the fire line wouldn't turn me into a forest firefighting expert. And during those weeks, many of my lessons had little to do with combustion and fuel.

When the crew arrived in Langlas Draw, after a long ride from Buffalo, Wyoming, to Meeker, Colorado, rumors circulated that we were putting out the blaze because Vice President Dick Cheney had a weekend fishing trip planned nearby. No mobile dining facilities were available to cater the fire camp, so the firefighters rode to the Sleepy Cat, a local lodge, for breakfast and a morning briefing.

"My people have gone a week without washing," one crew boss noted.

Portable shower facilities were committed to other fires, but the dirtiest crews could bathe in a nearby cabin.

"Jane and Mike Witt are really thankful that you saved their wedding site," said a member of the management crew, changing the subject. He held up a snapshot of the tuxedoed groom and his white-gowned bride in front of the firefighters.

Three hours later my crew stood at what we were told was the Witts' cabin, which looked like a set from a Ralph Lauren catalog—except for the foil firefighters had wrapped it in. For three days my crew hiked from the cabin to the pine and aspen forest above to mop up the fire.

Meanwhile, Secret Service agents arrived to prepare for the vice president's arrival and announced plans to ground the firefighting operation's helicopters.

"I'm not going to not be able to medevac a firefighter because the vice president's going fishing," the incident commander told me. He then pointed out to the agents how bad the publicity might be if that should occur, and they arranged to keep choppers working the fire. That Sunday three Black Hawk helicopters landed a mile and a half east of the fire camp to drop off Cheney and his fishing gear.

A few days later, at another lodge where fire crews dined, a boy stood on a picnic table and sang his thanks to firefighters for saving his home. When I commented on the boy's heartwarming song, a squad boss beside me scoffed. "He sang it better last year," he said, explaining that the boy's home had never been threatened by a fire, but his family owned a local lodge that counted on the "fire tourism" dollars the government spent on room, board, and supplies for wildfire crews.

My crew's most important lessons came on Storm King Mountain, where we hiked to a fire line stretched below what is known as Hell's Gate Ridge. Marble crosses marked the spot where each of the 14 firefighters who had died there fell. Skis, tools, toy fire trucks, crucifixes, yellowed photos, and sunbaked flowers decorated their bases. On top were coins and beer bottle caps from firefighters who, upon reaching the monuments, realized they had nothing else with which to pay their respects.

On July 6, 1994, 49 firefighters, including 16 smokejumpers and a 20-person team of Oregon hotshots, were on the mountain. To contain the puny but stubborn blaze, they built the fire line downhill toward a gulch—a dangerous practice when there's a fire on a facing slope. When a dry cold front blasted wind into western Colorado, the flames leapt onto a steep slope filled with highly combustible Gambel oak below the firefighters, who had never received the weather report predicting the wind shift.

The flames, ravenous for fuel, rolled up the 55-degree slope in waves. At its peak the blowup was running 35 feet a second up terrain so steep that the firefighters could hardly run at all. Chainsaws and gasoline cans exploded. Blasts of radiant heat seared skin, and superheated gases scorched lungs. The rocket-engine roar drowned out all but the closest shouts and screams.

Twelve smokejumpers and hotshots—America's firefighter athletes—perished. Two helicopter crewmen died trying to outrun the fire on the ridgeline where their chopper landed. The flames claimed 9 of the 20 Prineville Hotshots, from a town of 5,300 in the pine forests of central Oregon. For the first time America saw a tiny fire in the Rocky Mountains as a national problem. If this "nothing" of a blaze could cut the young heart out of a small town a full thousand miles away, the same could happen to any town in the country.

In the weeks that I was on the fire line, five firefighters died in western forest fires. Later in August a bus crash killed eight Oregon firefighters—bringing another tragedy to the land of the Prineville Hotshots.

Four days after I arrived home from my first firefighting trip west, another Connecticut crew was called up for a fire, this one in Montana. The national wildfire situation had hit Level 5, the highest federal preparedness level, drawing in all available resources. As the plane carrying the Connecticut crew approached Missoula, smoke from the burning forests obscured the airport. The flight was diverted to Great Falls, four hours away, where the crew boarded a school bus for the long ride to the Bitterroot Mountains.

7

THE BIGGER BLOWUP

IT WAS IN AN ALMOST INVISIBLE HOLE in the Bitterroots that America's war with wildfire was born. There, on August 20, 1910, "Big Ed" Pulaski, a ranger with the U.S. Forest Service, led 45 scruffy, un-trained firefighters to shelter in the Nicholson mine above the West Fork of Placer Creek when the "Big Blowup," a fire the size of Connecti-cut, threatened to incinerate them.

"The forests staggered, rocked, exploded and then shriveled under the holocaust," Betty Goodwin Spencer, an Idaho historian, recalled of the blaze. "Great red balls of fire rolled up the mountainsides. Crown fires, from one to 10 miles wide, streaked with yellow and purple and scarlet, raced through treetops 150 feet from the ground. Bloated bub-bles of gas burst murderously into forked and greedy flames . . . Fire brands the size of a man's arm were blasted down in the streets of towns 50 miles from the nearest fire line. The sun was completely obscured in Billings, 500 miles away from the main path of the fire . . . A 42 mile-an-hour gale swept Denver, enveloping it in a pall of smoke from the Idaho-Montana fires, 800 miles distant . . . You can't outrun wind and fire that are traveling 70 miles an hour. You can't hide when you are entirely sur-rounded by red-hot color. You can't see when it's pitch black in the af-ternoon. There were men who went stark raving crazy, men who flung themselves into the on-rushing flames, men who shot themselves."[1]

With Armageddon hot on his heals, Pulaski led his crew and two horses into the mine shaft to escape the inferno he described as having "the roar of a thousand freight trains." He kept the fire out of the pros-pector's hole with water he scooped from the shaft floor with his hat.

When some of his panicked firefighters tried to bolt, he pulled his pistol. "The next man who tries to leave the tunnel I will shoot," he told them.[2]

They all eventually fell unconscious as the inferno sucked the oxygen from the cave. Five never woke up. One survivor, finding Pulaski lying limp at the front of the shaft, announced that he was dead. "Like Hell he is" was his now legendary response.

Today the ranger's name rings out at every wildfire. The Pulaski, a combination axe and hoe that he invented after the ordeal, is the most common tool in the battle against forest fires, so even firefighters who don't know his story shout out his name on the fire line.

The ranger's legacy looms larger in the philosophy of firefighting that followed the blowup in the Bitterroots. Firefighters on the ground saw their efforts against the Big Blowup as a "complete failure." The fire killed at least 78 of the men fighting it, reduced much of Wallace, Idaho, to ash, and torched parts of half a dozen other towns. Mining camps, farms, and more than 3 million acres of timber burned.

But the fledgling Forest Service, just five years old and already hated in much of the West, chose to focus on the firefighters' heroic stand, rather than the futility of the battle. The American philosopher William James wrote of extinguishing wildfires as "the moral equivalent of war," suggesting that American youth be conscripted into an "army enlisted against nature."[3]

One of humanity's greatest allies was suddenly one of America's most reviled enemies.

Teddy Roosevelt was no longer president when the Big Blowup ignited, but he still led the nation's embryonic conservation movement. To the great hunter and adventurer, conservation was a hands-on affair. For western forests to flourish, they needed human help. Forest fires were not a weather phenomenon, but a wild beast the nation could exterminate to protect its precious timber resources, most majestic landscapes, and development of the West.

Gifford Pinchot, Roosevelt's friend, the founder of Yale's School of Forestry, and the first director of the U.S. Forest Service, had been pushed out of office by his political enemies when the great fire burned. But his philosophy of forestry, in which rangers actively managed the land for the good of both the timber industry and recreation, would continue to drive the agency. "Little G.P.s" — Pinchot's protégés — trained at Yale, then spread out through the forests of the West, where they imposed his philosophy for the next half century.

Pinchot, however, learned most of his forestry skills in Europe. He developed his vision of what the vast, untrammeled forests of the western United States could be by studying woodlands that, for centuries, had been inhabited by villages, crossed by roads, and managed for timber harvests. The management techniques used in woods long controlled by royalty often didn't fit in the vast, chaotic wildernesses of the West, or the messy democracy that controlled them.

Foresters debated two opposing options for managing fires in America's timberland. They could put them out fast or learn to live with them. They chose the former.

The next four directors of the U.S. Forest Service were all veterans of the fight against the Big Blowup in 1910, had all studied at Yale, and were all mentored by Pinchot. In 1935 the last of the Little G.P.s to serve as chief of the Forest Service, Ferdinand Augustus Silcox, instituted the "10 a.m. policy," under which the service was required to extinguish any wildfire by that hour of the morning after it was sighted. If they failed, the deadline was pushed back to 10 o'clock the next day.[4] With Silcox also helping to oversee the employment of millions of young men in the Civilian Conservation Corps and the Works Progress Administration during the Great Depression, he had such great manpower that his goal of "full suppression" of forest fires might have seemed achievable.

And then came Smokey Bear, star of the longest-running public service advertising campaign in history and a character as well-known as Mickey Mouse.[5] His message — "Only YOU can prevent forest fires" — brought the entire American public into the fight.

None of the firefighters with whom I clambered into a 40-foot school bus saw any connection between it and the mine shaft where Pulaski had held his own crew at gunpoint, despite their similar size and shape. But the Big Blowup started the war on wildfire in which we were foot soldiers a century later. It's a campaign that foresters and firefighters now realize they could never win. In fact, many of the battles in which they prevailed would just bring greater losses in the future.

PART III

THE CROWDED FOREST

8

MANSIONS IN THE SLUMS

A CENTURY AFTER THE BIG BLOWUP, America's fight against wildfire seemed like a victim of its own success.

As a nation, Americans have proved to be very capable forest firefighters. We still put out more than 98 percent of the country's wildfires during our initial attacks on them,[1] but the ones we can't snuff are bigger, hotter, faster, and more frequent than those we confronted before. By the time I was on the fire line, many foresters and firefighters believed that the blazes we couldn't stop had grown out of the ones that we did. Their message, in fact, became almost as codified into the western mythology of the twenty-first century as those of Pulaski, Silcox, and Smokey Bear during the twentieth: putting out all those fires but leaving behind the wood, grass, and scrub that otherwise would have burned overloaded our forests with fuel that was driving increasingly explosive fires.

While that was disastrously true in some forests and regions, in others the belief that every western woodland was an overgrown firetrap filled with sick trees was almost as misguided as the Forest Service's 10 a.m. policy. Sorting the reality from the myths required some walks deep into the woods, where the science was as nuanced as the landscape. Fortunately, I could find some of the answers in the parks and forests outside my own back door.

Historic photos and research by foresters show that before twentieth-century firefighters, foresters, miners, and ranchers changed the makeup of Colorado's ponderosa pine forests below about 8,000 feet in elevation, they typically held between 5 and 50 trees per acre, with parklike meadows between the trunks and their lowest branches 10 or 20 feet above the ground.[2] While some forests could be denser, in most ponderosa woodlands it was easy to get lost not because of how thick the forests were, but because they were so wide open a traveler could move unfettered in almost any direction.

Frequent fires creeping slow and low along the ground devoured scrub, deadfall, small trees, and low branches.[3] Ponderosas develop a thick, corklike bark that insulates them from flames, and many large, old ones have "cat face" fire scars and blackened trunks. The flames that marked them didn't hinder their growth into tall, majestic columns, but removed other vegetation that might have. I could still walk to a few ponderosa forests like that from my cabin in Colorado Chautauqua above Boulder.

After 100 years of the Forest Service extinguishing the ground fires that would have burned every few decades through most ponderosa pine forests, some of those woodlands — among the most iconic of the West — contain many times more woody fuel than they would naturally. Above Colorado Springs, Fort Collins, and Boulder, forests that historically had fewer than 100 trees per acre now have upwards of 500. In Arizona some forests have 40 times their natural load of trees: an acre that historically had 20 ponderosas is now crowded with 800 to 1,200.[4]

Forests in Colorado's Front Range have missed three, four, or five fire cycles that would have thinned them during the last century, Mike Battaglia, a young, slender researcher with the Forest Service's Rocky Mountain Research Station, told me as we headed into the woods above Fort Collins, where he's based.

"A forest, it's like a lawn," he said. "You have to cut your lawn every week."

In the ponderosas, fires were effectively the forests' lawn mowers, hedge clippers, and branch loppers. During the decades when the nation's firefighters put out every wildfire, the government had effectively fired nature's gardening crew.

Mike and I navigated through the jigsaw puzzle of public and private land that makes up Colorado's Front Range. Old mining claims sprinkled throughout national forest land now hold homes. Mike slowed the green Forest Service pickup to point out widely spaced trees that showed where fuel treatments — timber cut from the woods by hand or burned away by controlled fires — had thinned the national forest. The spaces closed back up when we passed into private land, where a century of fire exclusion had allowed the woods to grow thick, and nothing was done to remove the accumulating timber.

The transition between the two areas exposed a counterintuitive reality. While the public looks at wildfires as a challenge for the federal government, the National Fire Information Center reports that between 1993 and 2012 only 22 percent of them burned on federal land. More than three-quarters of these fires occurred on state or private land.

At a development called Glacier View Meadows we passed huge houses in woods so dense we could barely glimpse them. Blackened trees from a recent fire lined the ridge. Few of the homeowners had cleared enough trees to create defensible space — an area that will keep flames in the forest from reaching the home.

"Those people are so lucky they didn't burn," Mike said. "That's just ridiculous, how dense that is . . . It's irresponsible."

WHILE THE CANYONS ON COLORADO'S FRONT RANGE provide scenic and secluded homesites, some are more like slums of trees — overcrowded with evergreens struggling just to survive. But it isn't just firefighting that has left them that way.

Virtually all of our land uses impact the "fire regime" — how frequently, intensively, and expansively a landscape burns. When ranchers filled the West with cattle and sheep, the animals grazed down the grasses that carried the mellow ground fires that had thinned many ponderosa forests. Miners tramped the land around their claims to bare soil, with much the same effect. With nothing left to carry fire on the forest floor, trees that would have burned when they were small instead survived and crowded in. As the forests grew denser and darker, they provided a more hospitable environment for Douglas firs, which normally prefer north-facing slopes that are cooler and moister, but in this case followed the shade provided by the thickening stands. Mountain mahogany, Gambel oak, and juniper, which ignite easily and burn fast, pushed in beneath the pines and firs. Eventually a forest that once had trunks spaced hundreds of feet apart was filled with trees standing shoulder to shoulder, all of them fighting to get enough sun, food, and water. Undergrowth, fallen needles and leaves, and dead wood gathered around their feet. Diseases and pests such as the mountain pine beetle spread through crowded forests like the plague. Dwarf mistletoe infested many boughs, causing witches'-brooms of dry bristles that burned like haystacks.

Without the ground fires to burn off low branches, the growing pines developed "ladder fuels" — limbs running from the ground to the crown like rungs that fires on the forest floor climb into the treetops. In the canopy, the flames find abundant, easily ignited needles and are less sheltered from the wind, which pushes them to lean and leap onto other trees. The fire that would have burned slow and low on the floor of a forest of widely spaced pines now rips through the crowded canopy as an "active crown fire."

IN 1963 A. STARKER LEOPOLD oversaw a report on wildlife management for the National Park Service that called for reintroducing wildfire to federal woodlands. The son of naturalist Aldo Leopold, he grew up with conservation and wildfire. His father created the nation's first wilderness area in the fiery Gila Mountains of New Mexico and then died of a heart attack while fighting a grass fire on his neighbor's property in Wisconsin. Starker's report urged the Park Service to allow natural fires that didn't threaten development to run their course, and to set fires — prescribed burns — to thin the most overgrown forests. To many at the time, his ideas were the ravings of a madman.[5]

A quarter century later, when lightning fires that the Forest Service allowed to burn charred nearly a third of Yellowstone, politicians and the public responded with outrage. But by the time I was fighting fires, their benefit was clear. The forest ecosystems of Yellowstone were healthier a decade after the fires than they were a decade before them.

In 1995 the "Federal Wildland Fire Management Policy and Program Review" noted the need to put fire back on the land to correct what had become known as the "fire deficit" in the nation's woodlands. "The task before us — reintroducing fire — is both urgent and enormous," the report said. "Conditions on millions of acres of wildlands increase the probability of large, intense fires beyond any scale yet witnessed. These severe fires will in turn increase the risk to humans, to property, and to the land [with] which our social and economic well-being is so intimately intertwined."[6]

But the reintroduction of fire hasn't gone quite as planned. The excess timber and the warming and drying climate have made even carefully planned burns difficult to manage. And during the century in which the nation attempted to exclude fire from forests, they filled with homes.

AS PART OF THE GEORGE W. BUSH administration's Healthy Forests Initiative, the Forest Service prepared an ambitious plan of "fuel treatments" — thinning forests with axes and chainsaws and burning them with controlled fire to bring them back to health.[7]

But many woodland residents resisted efforts to bring flames back into the forests. More fuel meant more smoke and other potential impacts on human health. Prescribed burns could so reduce air quality that Colorado allowed burning on only a handful of days with enough breeze to dissipate the smoke, in order to minimize health effects on nearby residents, but not windy enough to blow the fires out of control. Others worried about the impact of controlled fires on wildlife and recreational opportunities. And some just didn't want flames in their forests.

In addition, the more desperately an overgrown forest needed fire, the more difficult a blaze set there would be to control. "We can't just put fire on the ground in some of these landscapes because they're too thick," Mike Battaglia said.

Agencies and politicians representing the various stakeholders stepped in, and the permitting process for burns grew as thick and difficult to navigate as the forests they were intended to heal.

The woodlands that were most overgrown were often those closest to homes and communities, where past fires were most aggressively snuffed. But those are also, of course, where the greatest resistance to prescribed burns and thinning projects lay. So controlled fires were often set in remote forests where they did little to mitigate the threat to development but were easier for the public and policy makers to accept. Research by Tania Schoennagel, a research scientist with the University of Colorado's geography department and Institute of Arctic and Alpine

Research, showed that between 2004 and 2008 only 11 percent of federal efforts to cut down or burn away dangerously overabundant forest fuels occurred in and around the wildland-urban interface, where homes abut fire-prone open space, despite government stipulations that significant resources to mitigate wildfires be focused on those areas.

Even projects that involve cutting down excess timber with chainsaws usually rely on a burn to remove the fuel. It's costly, and sometimes impossible, to carry out the wood left behind, so foresters usually plan to burn away the wood they cut down. But weather, air quality, and public resistance often shut down those burns, leaving piles of firewood stacked in the forest.

Across Colorado there are at least 180,000 "slash piles" of timber from thinning operations, waiting for incineration. Some have sat there for decades. And a forest thinned with axes and chainsaws but still filled with this debris is like a medical patient who doesn't finish a course of antibiotics — the treatment isn't complete. You may have moved the hazard around rather than eliminated it.

"If you just remove the fuel from the canopy and put it on the ground, you might not have a crown fire, but you've increased the severity of the ground fire," Mike said.

Many mountain residents view fuel treatments like surgery — a onetime fix. Prescribed burns, however, are more like medications for chronic illnesses that require regular dosage. If natural fires tend to burn a forest every 20 years, forest managers will need to introduce fire to the woods at roughly that interval to make up for wildfires they extinguish.

In the end, the treatment couldn't keep pace with the disease.

According to a 2006 report from the inspector general of the U.S. Department of Agriculture, which oversees the Forest Service, at the rate the United States applies fuel treatments, it would take 90 years to bring forests prone to unnaturally severe fires back to health.[8] To have any significant impact on fire risk, the Forest Service would need to treat 10 to 12 million acres a year, according to the 2005 "Quadrennial Fire and Fuel Review Report." That's nearly five times what the service currently treats.[9]

"Over the past 10 years we've consistently treated 2 to 3 million acres a year," Elizabeth Reinhardt, assistant director of fire and aviation management for the Forest Service, told me when I asked her about the nation's progress in healing sick forests. ". . . That's a huge chunk of the landscape, and I think we are seeing some payoff from that now."

That payoff, however, is small compared with the vast swaths of forests that need thinning. "When you look at [the] number of acres we burn every year and compare that to what we need to burn, it's pitiful," Mike told me.

In 2008 a statewide forest resource assessment showed that 6.8 million acres of Colorado's forested lands were significantly less healthy than they had been historically due largely to excluding fires from those woodlands.[10]

MORE THAN A CENTURY AFTER the Big Blowup inspired America to try to extinguish every wildfire in the country, foresters were finding that many of the nation's efforts to put fire back into the woods were just as misguided as that policy. Research published in 2012 by William Baker and Mark Williams at the University of Wyoming, who compared 100-year-old handwritten surveys from the U.S. General Land Office with data they gathered from trees along the lines the surveyors walked, showed that even in many ponderosa pine forests, dog-hair thickets (woods with a thick bristle of pines of the same age and size) and severe fires were more common than many foresters believed.[11] In fact, Baker and Williams argued, in many western forests the megafires foresters and firefighters saw as unprecedented responses to overgrown woodlands weren't out of keeping with historical fire regimes. Baker and Williams's research led to heated arguments at fire science conferences and in the pages of academic journals and the popular press. About the only thing the various sides in the squabble agreed on was that the United States was not dealing well with wildfires and it was going to see a lot more land burning.

During the time of this debate I walked with Tom Veblen, a geography professor who studies forests and runs the University of Colorado's Biogeography Lab, to look at the ponderosa pine forest at Heil Valley

Ranch, outside Boulder. Tom had just completed a project with Rose-mary Sherriff and a number of other researchers that showed, in Colorado's Front Range, just 16 percent of the forests below about 9,000 feet, where ponderosas are the dominant species, had increased fire severity due to fires extinguished in the past leading to overly thick growth. Other forests had a history of large, intense fires even before the U.S. government started snuffing every wildfire.[12]

Sixteen percent is not an insignificant number, however, as Tom noted. And the areas showing an increase in wildfires over the past century are precisely where the most mountain homes have been built. The lesson of the new research by Sherriff and colleagues isn't to stop thinning and burning the woods, but to focus those treatments where they will do the most good — overgrown ponderosa forests adjacent to homes and other development. Of course, those are the areas where resistance to such treatments is the greatest.

The remaining 84 percent of Colorado's Front Range includes many other forest types — high country dominated by lodgepole pine, sub-alpine fir, and spruce forests, arid mesas of piñons and junipers, and even ponderosas mixed with other conifers higher in the mountains. In those forests increases in the size and severity of wildfires don't correlate with an increased number of trees due to fires being extinguished. Most of the higher-elevation forests only burn every 100 to 300 years, so the nation's century of fire suppression could only have interrupted their natural fire cycle once, if at all. Many of them always had big, intense wildfires.

WHILE THE DEBATE RAGED ABOUT which forests were seeing larger, more severe wildfires after too many fires were snuffed in the past, other scientists argued about whether wildfires were always a bad thing. Humans don't want intense blazes near their homes, but other plants and animals depend on them.

Even the biggest and hottest fires leave pockets of slightly burnt or unburnt vegetation. Foresters refer to that pattern of lower- and higher-intensity burns as a "mosaic." The right mix leads to a healthier forest with a variety of habitats for animals, more diverse vegetation, and a cleaner and more efficient watershed. In moderately burnt areas seedlings can grow faster than they would have if the fire had not come through at all, while areas cooked by high-intensity fires may not show green for years. The most intense fires destroy both the trees and their seed stock, and sometimes leave the soil "sterilized," with few nutrients available to support new trees. But even bad burns are important parts of the mosaic.

"Snag forests" — remnants of severe fires in which only a few burnt trunks still stand — are a critical habitat for many species. Black-backed woodpeckers, for instance, are dependent on the charred trees in intensely burnt forests for their nests and the insects they eat.[13] In the Northwest, reductions in the amount of severely burnt forests over the past century led to declines in the woodpecker's population. According to Dick Hutto, who has conducted intensive research on the woodpecker and other birds that thrive only in severely burnt woodlands, that would imply that recent intensely burning wildfires aren't as unnatural or unhealthy as many perceive them to be.

In the Southeast, longleaf pine forests that grew thick when fires were removed from the landscape a century ago showed similar reductions in bobwhite quail numbers. Hunters of the birds drove a movement to reintroduce fire in Florida and Georgia forests.

FOR MOST RESIDENTS OF COLORADO, human and otherwise, the greatest threat from wildfires isn't the flames. The 2002 Hayman Fire scorched nearly 140,000 acres of the Upper South Platte watershed, which provides Denver's million and a half residents with more than 75 percent of their water.

For two days in 2012, as wildfires once again overtook the state, scientists, land managers, and concerned citizens filled the city's REI store to talk about the fire 10 years earlier that they were still recovering from. The parallels were impossible to miss.

"This year looks very similar to 2002," Don Kennedy, an environmental scientist with Denver Water, said of the drought overtaking the state.

Runoff from the burn scars of the Hayman Fire and the 1996 Buffalo Creek Fire continued to fill the Strontia Springs Reservoir with ash, mud, cinders, and debris from the fire, Kennedy told the audience. "Engineers, when they built the reservoir, estimated that there would be

20,000 cubic yards of material per year," he said. "We got 200,000 in one rain event."

The sediment that washed down from the burn scars was heavy with cadmium, copper, and lead, which can poison the drinking water supply and are difficult to filter out. While the water showed somewhat higher concentrations of all three toxins, it was still below federal standards for drinking water. The sediment that slid into the water, however, had levels as much as 80 times higher. After other fires, dangerous levels of arsenic, mercury, zinc, and cyanide found their way into public water supplies. Denver Water spent $26 million digging toxic sludge from the Hayman Fire out of its reservoirs and managed to remove only about a quarter of the 625,000 cubic yards of fire-contaminated muck, forcing the utility to spend even more money treating the water.[14]

Denver Water realized that its money would be better spent preventing high-intensity wildfires in the first place. Between 2012 and 2015, it planned to spend $16.5 million to fund thinning projects and prescribed fires on 38,000 acres of land. One of those projects was in the Lower North Fork watershed, about 30 miles west of Denver.

9

THE BLACKLINE

Conifer, Colorado — March 19, 2012

THE MORNING AFTER THE HEARTSTRONG FIRE, as doctors in Greeley fought to save Jennifer Struckmeyer's life, 100 miles to the south the Colorado State Forest Service prepared to treat a sick mountain.

Trees so overcrowded the forests surrounding the Lower North Fork of the South Platte River that they risked fueling a fire big and hot enough to threaten both the woodlands and the watershed. The CSFS determined that a prescribed burn set by firefighters to remove excess timber, scrub, and grass was the best way to bring the woodlands back to health.

That Monday, Kirk Will, "burn boss" for the fire the CSFS planned to set three days later, stood in the woods watching crews "blacklining." Firefighters sprinkled a flaming mix of diesel fuel and gasoline onto the ground from drip torches that looked like watering cans with spouts of fire. From above, the arcing line they blackened on the forest floor looked like it had been drawn with a skyscraper-sized Magic Marker. According to Will, the line was between 66 and 132 feet wide.[1] After the fire, however, a report would claim that it was about 50 feet wide,[2] while a map of the line identified it as just 10 to 20 feet wide.[3] A week later, the width of the blackline would be just one among many details about the burn that didn't add up.

The blackline — ground burned void of anything flammable — encircled the north and east sides of a plot of forest owned by Denver Water on the eastern edge of the Rocky Mountains about 30 miles southwest of

Denver. The south side of the plot was contained by a "handline," which was effectively a trail firefighters dug down to mineral soil with rakes, shovels, hoes, and axes to remove anything flammable so that a fire on the ground couldn't pass it. The blackline and handline connected with a road on a ridge to form a continuous perimeter — a "control line" intended to hold the fire in a plot known as Unit 4 like a corral holds wild horses.[4]

Inside the perimeter were 50 acres of masticated fuels — timber cut down and chewed up with chippers and chainsaws four years earlier.[5] The chunks of ponderosa pine and Douglas fir, some the size of a man's leg, were spread shin-deep below the pines and firs that remained standing.

Building the control line had taken the better part of a year, so it was satisfying to watch the circle close. Like every other aspect of a prescribed burn, it was a labor-intensive, painstaking process. And Unit 4 was just one of six areas in the Lower North Fork that the CSFS planned to burn over a five-year period.

The previous spring Kevin Michalak, an engine boss and technician with the CSFS, had lit a test fire to see if the prescribed fire intended to heal the forest might accidentally destroy it. The test hadn't gone well. The small blaze had jumped the line intended to contain it.[6] Nonetheless, the prescribed burn planned for Unit 4 went forward.

When the blackline was completed, Will and his crew "mopped up," the firefighter's term for making sure every ember was out cold — soaked, crushed, or buried. The following day, Tuesday, a crew put hoses, a motorized pump, and a portable water tank into position. The water would be a critical tool for mopping up the burn and extinguishing any "spot fires" that might ignite outside the line. Wednesday Will confirmed with 10 different agencies and fire departments that their promised manpower and equipment would be there the next day, when they lit the fire inside the control line and let it run.[7]

At 8:30 on Thursday morning more than 80 people, twice what was called for in the burn plan, gathered for an hourlong briefing at Jefferson County's Reynolds Park, where the firefighters could look out at the mountains where they would be working. Then they drove a few miles south, unlocked the gate to the Denver Water land, and headed to Unit

4. A 20-man firefighting crew made up of inmates from the Colorado Department of Corrections joined two 20-person crews from the CSFS to make sure the blaze stayed within the control line. Six engines from as far away as Fort Collins surrounded the unit, with two more standing by. Two people set up air quality monitors. Rocco Snart, fire management officer with the Jefferson County Sheriff's Office, was the safety officer and was also overseeing the training of eight firefighters who were refreshing their wildland firefighting credentials.[8] Kevin Michalak was in charge of mopping up. Also overseeing the fire was Allen Gallamore, district forester with the CSFS. Six members of the Platte Canyon Fire Protection District crew would light the blaze.[9]

The state's prescribed burn plan had more than a dozen pages of questions for Kirk Will to answer regarding the controlled fire. It included a "complexity analysis" of how difficult the fire would be to ignite and manage; current and predicted weather conditions; smoke management and air quality; how the work was funded, scheduled, and staffed; and a plan for what to do should the controlled fire escape its containment and turn into a wildfire. Its final "Go/No Go Checklist" asked: "Has the burn unit experienced unusual drought conditions or does it contain above normal fuel loadings which were not considered in the prescription development?"

The Haines Index that the CSFS relies on to measure drought showed very low moisture at the burn site.[10] Other measures showed an even more dramatic drought.

Snowcapped mountains trickle water into the forests below, thus keeping woodlands moist for months after the snow stops falling. In 2011, when the Lower North Fork prescribed burn was being planned, Colorado had extraordinarily heavy snowfall. But a year later, when the state foresters were going over their checklist, the state was at the end of one of its driest winters on record. At the time of the burn Colorado's snowpack was about half the annual average, and NOAA reported it was the driest March in state history. By the end of May, 100 percent of the state would be in drought.[11]

The dead timber spread over the forest floor was certainly a far heavier fuel load than normal, but one the firefighters believed they understood.

On-site weather observations noted a maximum temperature of 59 degrees, with relative humidity of 21 percent and winds running east to northeast up to 9 miles per hour. There was enough of a breeze to dissipate the smoke, but the winds wouldn't be strong enough to blow up the fire. Cool temperatures would dampen the flames. Conditions seemed ideal for a prescribed burn, even if there was barely a fraction of the normal amount of snow on the ground.

So despite the thick load of broken timber piled on the forest floor and unusually warm and dry weather, Will noted no unusual drought or fuel load in the Go/No Go paperwork.[12]

The burn boss went over the plan with his firing boss, who would oversee the crew lighting the fire, and the holding specialist, who would lead the firefighters charged with keeping it corralled. They agreed that each item was a go. At 11:29 Will signed the checklist. The firing crew lit a test fire, and at 11:45 Will noted it went well. The fuel was dry enough to burn, and weather conditions would make the blaze easy to contain. At noon firefighters began dripping fire onto the forest floor.[13]

A WEEK EARLIER FIREFIGHTERS HAD HUNG SIGNS along the roads surrounding the forest warning of the pending burn, but Kim Olson hadn't seen any of them. Neither she nor any of her neighbors, a little more than a mile from Unit 4, were aware of the planned fire.

A petite dancer with short but wild blond hair, Kim looked only a few years older than the university students she sometimes taught, but she was actually the mother of three children. She and her husband, Doug Gulick, had moved into their home on Kuehster Road, atop a wide, flat ridge of forests and meadows, in the mid-1990s. They'd seen a lot of other homes sprout up in their community since then.

The daughter of a firefighter, Kim was usually the first person in the neighborhood to smell any smoke, the woman who was always looking for smudges in the sky. Even her family had begun to see her wariness as bordering on paranoia.

Although no scent, sight, or sign alerted her to the planned burn, while the firefighters were igniting the forest, Kim headed to town to purchase fire-escape ladders for the family's home. She couldn't say why.

10

SLOP-OVER

Conifer, Colorado — March 26, 2012

IF ANYONE UNDERSTOOD THE NEED to burn the forests around the Lower North Fork, it was the man who had just taken charge of extinguishing fires there. Bill McLaughlin was five weeks into his new job as the chief of the Elk Creek Fire Department when the members of the Colorado State Forest Service lit their torches in the forest. His office was about five miles from the prescribed burn on the Denver Water land.

When McLaughlin interviewed for the job, he knew he was stepping into a troubled department. Two previous chiefs and the department's board president had all resigned amid scandals and turmoil. And like many volunteer departments (the chief was paid, but most of the firefighters were not), Elk Creek's budget was stressed. It hadn't had a levy increase since the 1950s. Although an explosion of development had increased the fire department's budget — Jefferson County was the fastest-growing county in Colorado — it didn't come close to keeping up with the department's needs. Radios were old and couldn't be used to talk with many other fire departments and sheriff's offices, or with the Colorado State Forest Service. Outdated maps, some of them decades old, didn't show most of the new homes, developments, and roads.

"Realistically, with the resources that we've got, we can handle a five-acre fire in a low-complexity situation, without a lot of wind or immediate threat to homes," McLaughlin told me. "That's the limit. It gets beyond that and we have to rely on outside assistance."

Some of the forests he'd toured during his interviews were more stressed than the department's budget. Many were overgrown dog-hair thickets after decades in which firefighters had extinguished every natural blaze.

At higher elevations they'd come to vast mixed-conifer and lodgepole pine forests that naturally burn in what are known as "stand-replacing fires," which kill almost every tree on entire mountainsides. He could see that the trees were all about the same height and width, a sign that they had sprouted at the same time and grown up together after a fire incinerated all their predecessors. He knew that was how these forests regenerated: lodgepoles have cones that release their seeds in a fire and grow well on recently burnt ground. He also knew that they tended to burn big.

McLaughlin found something else that crowded the woods and would be far more complicating to his job — homes.

Along Elk Creek, the Bucksnort Saloon epitomizes the neighborhoods McLaughlin would defend if he took the job: hard to get to, hard to leave, and smack-dab in the middle of a natural chimney. A narrow road follows the creek through the bottom of a canyon to the log-walled tavern. Bikers and tourists put their feet up on its deck, which overhangs the rapids of the creek, and the ceiling is covered with dollar bills bearing messages from visitors from around the world. Thick forests crowd both sides of the gorge that snakes miles in both directions from the saloon. Ancient, tumbledown mining shacks hang from cliff tops, while new mansions rise in clearings in the dense, dark woods. The winding road is so tight that occasionally drivers have to pull off to the side to allow oncoming traffic to pass. McLaughlin couldn't imagine how he would get a fire truck through the narrow lanes, how he would defend those homes with a fire filling the canyon, or how he would evacuate the saloon with traffic clogging the road.

"This is a community that is at extreme risk for wildfire," McLaughlin told the board when he got back to the department. "We have literally thousands of homes scattered around out here, most of them down one-lane dirt roads — one way in and out, a lot of fuels, a lot of forest . . . that we expect to burn with crown fires. And people are living right in the middle of it."

• • •

CALM AND QUIET, WITH GRAYING HAIR and a soft, roundish face, McLaughlin looks more like an executive than a mountain firefighter. He grew up in New Jersey, where his father was stationed in the U.S. Army.

"I stayed as long as I could stand it and then I drove as far as there was a road — literally," he told me. "Homer, Alaska — the nickname there is 'the end of the road.' I got on a boat there to Kodiak. And then I got into a bush plane and I flew to the far end of the island. And then I felt like I was away from New Jersey. I was 21."

Hungry for adventure, service, and the kind of camaraderie he'd grown up around in an army family, McLaughlin joined a volunteer fire department in Alaska. "My second call ever we had two fatalities," he recalled. ". . . A commuter plane crash . . . I was still in rookie class."

Over the next decade he moved from volunteer to professional firefighting and from Alaska to Washington State, where he was the fire chief at Friday Harbor, in the San Juan Islands. There McLaughlin was not only a firefighter but a fire setter. He worked on prescribed burns to restore the Garry oak savanna on the islands. The flames burned away snowberry and thinned out stands of Douglas fir, both of which outcompete the native oaks in the absence of fire. "It's about time," area residents would tell him.

He knew that it would be much more difficult to put fire back on the ground in Colorado. The forests were not only overgrown but were far drier than they were in western Washington. And as he looked at the developments, businesses, and homes sprinkled throughout them, he knew that the local population would be far less accepting of controlled fires than the population in the San Juans had been. One way or another, however, the forest was going to burn.

"If we don't have fires run through this landscape, we're going to continue to change the makeup of the forest," he said. "And it is going to continue to [have] heavier fuel loading and move more towards a less often but much more destructive fire regime."

McLaughlin took the job knowing that helping out with prescribed fires would be part of his job and that if one got away, his crews would be fighting it. A month before the Lower North Fork prescribed burn, the Elk Creek Fire Department worked with the U.S. Forest Service on a burn of enough slash from thinning projects to fill 18 tractor-trailers.

"They actually hired me specifically to ... bring the level of pre-paredness both for the department and for the community up to where it needs to be," he told me.

McLaughlin had been in his new job for four weeks when the Colo-rado State Forest Service called to let the department know about the prescribed burn they planned to ignite on the Denver Water land on Thursday, March 22. Even though Colorado was in the midst of its dri-est March in history and a record warm spell, McLaughlin wasn't con-cerned about the plan to light the woods on fire. The CSFS hadn't pro-vided him with a copy of the burn plan, so he didn't know enough about it to get worried. He just knew that the forest was as thirsty for fire as it was for water.

KEVIN MICHALAK WAS IN CHARGE of mopping up after the pre-scribed burn in the Lower North Fork — making sure every last ember was out. He was pleased with how the burn had progressed, and confi-dent that it was out cold when he left it Thursday night. But he couldn't say that it had gone like clockwork.

Variable winds had caused the flames in Unit 4 to jump around more than he would have liked, and a few firefighters had fretted that the in-terior of the unit had burned too hot.[1] Embers rolling down hillsides and across the control line also had ignited four or five small spot fires. Tracking down tiny smokes outside the prescribed burn area is like hunting down tiny, burrowing animals before they can reproduce and climb into the treetops. That wasn't hard with some 80 firefighters on the scene. Engines and crews on foot quickly snuffed any smoke or em-ber they spotted outside the line.[2]

They mopped up a swath about 100 feet wide between the control line and the interior of the burn.[3] Inside that perimeter, a few stumps still smoked, but the crew was confident there was no way for anything burning to escape. Still, the thick layer of chewed-up trees covering the forest floor like a giant campfire was a fuel bed that few of them had much experience with. They'd have to keep an eye on it through the weekend.

Michalak returned to the burn site around 9 a.m. Friday to complete the mop-up. To finish snuffing the blaze, he had only a fraction of the manpower that had been there to light it — a crew of prison firefighters, four engines, and an all-terrain vehicle that could carry water through the unit. Heavy smoke rose from a couple of spots deep inside the burn, and one spot fire crawled outside the control line before they smothered it. They threw smoking branches near the perimeter deep into the burn area, dug up smoldering roots, and soaked the edge of the burn.[4]

The burn plan's mop-up and patrol guidelines were based on the Keetch-Byram Drought Index, a measure of how dry, and consequently how combustible, a landscape is. It led them to devise a 200-foot-wide perimeter of "cold black" — ground burned, scraped, and soaked with water into something that looks like a black tar racetrack. Michalak reported achieving that goal by the end of the shift.[5]

By the time he left, however, the sky seemed to be inhaling in preparation to blow hard. A ridge of high pressure rose between New Mexico and South Dakota, while a low-pressure trough formed along the Pacific coast. Temperatures increased nearly 20 degrees during the weekend after the burn — Denver set a temperature record — and relative humidity plummeted into the single digits.[6]

Saturday morning — two days after the burn — Michalak went back to the burn and traveled along the western and southern edges of the unit on foot and in an ATV. The fire was lifeless, with just two wisps of smoke rising from needles in the center of the charred woods. He noted that there was no "heat" within 200 feet of the unit's perimeter. The entire blackline was cold. It seemed impossible that an ember deep in the burn could reach the forest outside it.[7] He left the unit early that afternoon and contacted Kirk Will, the burn boss, and the district forester, Allen Gallamore.

The burn plan required patrol of the site for at least three days after the fire to make sure it stayed out, but Michalak and Will were so confident it wasn't a threat, they decided to leave the burnt forest unattended on Sunday, March 25.[8] Michalak wouldn't return to it until Monday morning.

At 5:26 on Sunday morning the National Weather Service issued a

Fire Weather Watch for Monday, calling for "strong winds and low relative humidity."

At 12:15 p.m. the National Weather Service upgraded that forecast to a Red Flag Warning calling for very low humidity, sustained winds between 22 and 32 miles per hour, and gusts to 60 miles per hour. A breakdown of the upper ridge in the atmosphere would move the high-pressure system east. Behind it, a cold front would move across Utah and into Colorado, pushing high winds, low humidity, and high temperatures before it. A vertical lift between the fronts would function like a vacuum on the mountains, creating a highly unstable atmosphere. Conditions like these drove Colorado's most notorious fires — the killer South Canyon Fire on Storm King Mountain, the massive Hayman Fire, the Bobcat Gulch Fire, the Overland Fire.

Michalak arrived at the unit by 10:30 Monday morning, March 26, half an hour after the Red Flag Warning went into effect. The sky was cloudy, and it had been cool enough overnight that he didn't see too much to worry about with the weather. He planned to pack up all the firefighting equipment and brought only a pickup and an all-terrain vehicle rather than a fire engine.[9]

Two CSFS firefighters, Rob Kriegbaum and Ryan Cox, had come along to help roll up hoses and map the perimeter with a GPS. The only smoke they noticed was from a scorched log and the pile of needles Michalak had seen smoldering two days before. Both were deep in the black. By 12:30 they had loaded the gear into the truck and the ATV. Michalak noticed the wind shift to the southwest and increase to about 15 miles per hour. The firefighters drove to the north side of the unit, where embers picked up by the southerly wind might land outside the black in the unburnt forest. The wind gusted to 20 miles per hour. Embers jumped "like fleas" toward the control line. Michalak noted several small puffs of smoke from the duff, needles, and charred wood inside the unit. Kriegbaum and Cox grabbed hoes and rakes to snuff the "duffers" that were springing up like leaks in the charred forest floor. Michalak could see they were going to need water, so he drove the ATV to a creek to fill up its 70-gallon tank. He returned 15 minutes later to find the two firefighters swinging their hand tools at two small spot fires that had broken out in the green forest about five feet on the other side of the road. When the spot fires were dead and buried, the three firefighters headed back into the black.[10]

The flames resurrected by the relentless wind were as dumbfounding to the firefighters as corpses rising from graves. The thick bed of crushed wood hid hundreds of embers. Michalak was quickly exhausting his water. Embers he drowned dried out and reignited within 15 minutes. The growing wind ignited small blazes faster than the firefighters could snuff them. Both Michalak and Kriegbaum called Kirk Will, the burn boss, requesting an engine to help.

They ran out of water less than an hour after filling up, leaving them with only hand tools to attack the small flames. Michalak noticed thickening smoke on the south side of the unit but found only duffs smoldering well within the blackline. The crew on Colorado State Forest Service Engine 862 arrived, left their engine for Michalak, and headed back to their office.[11]

After they left, Michalak looked back south and saw a thick plume of smoke. Driving the ATV, he found a spot fire no bigger than a quarter acre in size. Just north of the spot fire, Michalak saw an old handline that firefighters had scratched into the hill the previous fall. It was a reminder of the challenge he now faced.[12]

Five months earlier, in October 2011, the day after he and his crew had burned a portion of the blackline intended to contain the current fire, an ember had ignited a spot fire. The handline firefighters had used to round up that earlier escaped blaze was still visible, emphasizing how difficult it was to control a fire in this forest. It marked one of at least three fires set in preparation for the current burn that had sparked spot fires that had nearly gotten away.[13]

The burns in the Lower North Fork, in fact, had repeatedly escaped prior to March 2012, but they had been corralled and killed before anyone outside the CSFS heard anything about them. The area was well-known to firemen and area residents to channel gusty and unpredictable winds. Investigators of the escaped fire that Michalak was fighting to contain would find not one but several ignition sources outside the control line.[14] A rain of embers had escaped the controlled burn to ignite a wildfire.

With just hoes and shovels, and no water, the three firefighters stood no chance of stopping a fire in chopped-up timber piled in a green forest. They gathered their gear into a staging area and waited for reinforcements. What would soon be known as the Lower North Fork Fire was pushing hard to the north like a herd of smoky beasts.

Michalak reported the spot fire to Kirk Will and requested four more engines, two crews of firefighters, and another officer to oversee the effort. He also asked for a Jefferson County Type 3 Incident Management Team — a first response to a forest fire.[15] Jefferson County 9-1-1 Dispatch called the Elk Creek and North Fork fire departments.

The term the firefighters broadcast over the radio — "slop-over" — sounds about as problematic as beer sloshing from a mug. But every firefighter who hears the phrase recognizes the threat behind it: a controlled fire is trying to run wild.

CHIEF McLAUGHLIN WAS SURPRISED to see Monday morning's forecast predict winds even stronger than the Red Flag Warning called for. He filled up his canteen and called in a few more volunteer firefighters.

The department had already responded to two small wildfires that week, so on top of trying to make headway on the long-term and endemic problems his department faced, the new chief was also racing to get the volunteers ready for what was clearly going to be an early and active wildfire season. Tests of fuel moisture and long-term weather forecasts predicted a much drier spring than usual.

Chinook winds gust every spring in Colorado. They originate in the Pacific, drop their moisture on the west side of the Rockies, and then blow warm and fast down the eastern slopes. The chinooks in 2012 were unusually strong. They normally hit while the mountains are still covered in snow, which shelters fallen timber, scrub, and grasses from the gusts that can dry them out and push flames onto them. Runoff from the gradually melting mountain snowpack keeps the forests moist and "greens up" the vegetation, making it less likely to burn. But in 2012, March — usually the state's snowiest month — saw almost nothing white in the air or on the ground. On average more than two inches of precipitation falls in that month around the Lower North Fork. In 2012 only half an inch fell, and that had come 20 days before the burn.[16] The mountains were tan rather than white and green.

"Looking back historically, spring was not considered part of fire season in Colorado until the very recent past," McLaughlin told me. "It's been largely the last decade that they've seen those spring fires occurring."

Across the West during the previous decade, mountain snowpack had melted off weeks, and sometimes months, earlier than it had during the previous century. "Fire season around here was usually a couple of very short windows," McLaughlin explained. "Those windows just keep expanding further and further."

The winds started blasting shortly after ten that morning, and the department's first call came not long afterward. A gust knocked down a power line into a grassy field in Aspen Park, a suburb 30 miles from Denver. A resident from one of the nearby homes was spraying water on the blaze with his garden hose when the Elk Creek crew arrived and put it out.

McLaughlin was getting ready to leave when, at 1:55, the call came for help extinguishing a one-acre slop-over from the prescribed burn on the Denver Water land. And with that alarm the hazard he hoped to mitigate in the woods, the weaknesses he hoped to strengthen in his department, and the disaster he hoped to prevent seemed to cascade onto him.

The call was for a single fire engine, but the chief sent two, along with a water tender — a truck with a tank for others to draw from. But he drove past the turnoff into the property twice. It wasn't marked on his old map.

"The technology exists so that we can have maps that show topography, show vegetation, that will let us actually predict where fire spread will occur, that will even, when we bring up somebody's address, . . . look at evacuation routes," he said. But that technology was just an expensive dream at the Elk Creek Fire Department.

"The maps we use to just find somebody's medical call are just . . . a road map that someone's copied," he said. "They don't have addresses written on them, no houses. Basically about as primitive as it gets.

They're not even as nice as the typical road map you would pick up if you went into a gas station."

Northwest of the Denver Water land, subdivisions like Aspen Park and Conifer Meadows were named for trees but filled with homes. Many of the developments were not on the chief's out-of-date maps, and he wondered how he could protect homes he couldn't even find. Some homes were in a no-man's-land that was not part of any fire protection district. Many firefighters had taken to relying on their smartphones for guidance, but having a map that vanishes when a cellphone's connection fades could be worse than having one that is obsolete.

At 2:11 dispatch reported that the fire was one acre in size, "with low spread potential," but by the time McLaughlin arrived, it was five acres and growing fast. Michalak, the incident commander in charge of the mop-up, was waging a losing battle to contain the slop-over. Curt Rogers, the chief of the North Fork Volunteer Fire Department, arrived a few minutes before McLaughlin with two fire trucks, along with firefighters from the Jefferson County Sheriff's Office.

By then, winds were gusting to more than 60 miles per hour. Firefighters who already had their hands full had to use them to hold their helmets on. Others sheltered behind fire trucks. Blizzards of embers blew sideways, blasting firefighters trying to contain the blaze. When Chief Rogers held up his handheld weather meter, a gust buried the needle at its maximum — 80 miles per hour.

THE GROUND FIRE WAS INCHING DOWNHILL to the northeast like molten lava. Flames a foot high filled the forest floor and leapt as high as eight feet, torching individual pines. "It looked like just a four-inch bed of charcoal burning everywhere," McLaughlin said.

When Rich Palestro, an engine boss with the CSFS, arrived at 3:30, he guessed the fire had grown to 20 acres. He staged his truck at the south end of the fire and joined a crew setting up a pump and hose to try to flank it, but they retreated when he saw the flames racing fast enough to overrun them.[17]

"We'll be hitting some trigger points pretty soon here," Rogers radioed to Jefferson County 9-1-1 Dispatch at 2:38. "I'm sure we'll be having some evacuations pretty quick."

Rogers's certainty was misplaced. Outdated technology and confused communications — other items high on Chief McLaughlin's fix-it list — got in the way.

McLaughlin called dispatch two minutes later to request more reinforcements. "We need to start looking at evacuating to Reynolds [Park] and potentially further up," he told them.

The sheriff's office runs evacuations, but usually waits for a fire's incident commander to request one. Michalak, with the CSFS, was still commander of the fire, but he couldn't hear McLaughlin's and Rogers's missives to start getting people out of the way of the blaze. The chiefs didn't know which radio channels the CSFS was using. Jefferson County dispatch tried to patch McLaughlin through to Michalak but failed. "If I wanted to talk to the incident commander, I had to go down there and track him down," McLaughlin said.

As other departments were called in to help, the communication problems grew. Many departments used 800 band radios, which have greater range but work poorly in mountains and canyons. Departments that work rugged topography usually stick with older VHF technology. The next fire department north of McLaughlin's used UHF radios, with yet another set of frequencies. The three systems don't talk to one another. Even when departments do use the same radio systems, each communicates on its own channel until a common one is established among them.

As the fire grew, firefighters began using cellphones to contact colleagues just a few hundred feet away. McLaughlin was working four channels on two radios and was running back and forth between other firefighters whom he couldn't raise on the two-ways.

"We found fire engines that had come in to help out that had never talked to anybody," he said. "They were just driving around doing whatever they felt they needed to do because they didn't have the channels to communicate on."

Help, unsurprisingly, was slow to come.

Colorado is one of just two U.S. states with no state fire marshal to order resources between jurisdictions. Decisions about which departments send aid to fires in other districts are made at the local and county levels. One fire department asks for assistance from another as "mutual aid" but can't be sure how much of the requested help will arrive. "It's . . . a much more cumbersome and slow approach," McLaughlin said.

When the alarm went out for the slop-over in the Lower North Fork, smokes were sprouting from Colorado Springs to Fort Collins. Some departments couldn't respond to McLaughlin's call for help because they had wildfires in their own districts or wanted to keep resources close just in case. Others may have been worried about the costs. Although Colorado has an emergency wildfire fund, assisting departments don't know if they'll get any of that money, or how much, until weeks or months after they respond to a fire.

"We don't know if the state will end up picking up the tab or not," Chief McLaughlin said. "I think that's hindering some departments from providing the assistance levels that they could."

THE CHIEFS WATCHED THE FIRE slowly descend toward a drainage. On the other side was a south-facing hillside thick with scrub oaks, junipers, and oily, drought-stressed brush.

Fire responds to terrain in the opposite way that humans do — running far faster uphill than down. The incline, sunny southern exposure, and highly flammable vegetation on that hill would make the fire sprint up the slope it was approaching.

"If we can stop it down by the drainage, we'll be doing good," Rogers radioed at 3:26 p.m. "But if it spots across to the other side, it's off to the races."

Rogers estimated that it would take two hours for the fire to climb the long slope. At the top were the homes of Pleasant Park and Kuehster Road.

11

OFF TO THE RACES

KIM OLSON — THE FIREFIGHTER'S DAUGHTER — and her husband, Doug Gulick, had bought their property on Kuehster Road in 1997. "We made a hunting cabin into a house," she said.

The soil along the road, unusually rich for its 8,000-foot elevation, had drawn the Kuehster family to build a farm there in 1810. Two hundred years later, it was the solitude, verdant forest, and stunning views — Pikes Peak to the south, Mount Evans to the west, and the high plains spreading east past Denver — that led people to construct homes among the trees. The rocky pulpit of Devils Head stabs the sky from the Rampart Range, and the towering Cathedral Spires overlook Elk Creek. The good and evil labeling made Kuehster Road seem that much more like heaven.

The neighborhood also included all the points of Colorado's polarized social and political compass. Buddhist prayer flags with wind horses and snow lions hung from one porch, a banner with a rattlesnake warning "Don't Tread on Me" from another. Wind chimes rang from some properties, rifle shots from others. Bulldozers were parked next to corrals of horses. Somehow everyone seemed to get along.

"Everybody likes their space," Kim Olson said, "but there's never been a time when somebody didn't help somebody off the road or lend them firewood, or a tool, or a cup of flour."

Ties revealed themselves in surprising ways. Kim's husband worked with a thermal analyst named Sam Lucas at Lockheed Martin, a few miles away. His father was also named Sam and lived near Kim and Doug with his wife, Moaneti. Sam the elder was a retired mechanical engineer who chopped wood for his stove and taught Sunday school. Moaneti "prayed about everything, including parking spaces at Walmart," Jack McCullough, of Red Rocks Fellowship, recalled. She painted landscapes, raised chickens and vegetables, and sang in the church choir.[1]

The high school sweethearts had been married for 58 years. When they were newlyweds, Sam hitchhiked home from college to see his wife, then worked two jobs so she could stay home with their children. They thought of themselves as "mountain people" and didn't see the land where they eventually built their dream home on Eagle Vista Drive until the thigh-deep snow that covered it when they bought it melted away.

Often hardships in the community turned into social gatherings. Some neighbors would see one another only when a blizzard hit. They'd sit on their plows or lean against their shovels and catch up on grand-kids and work and summer plans as the snow piled up around them, then get back to clearing the road and one another's driveways.

Andy and Jeanie Hoover provided the most popular venue for neighborhood gatherings and were renowned for their annual Christmas party. Andy — the big, jovial grandson of President Herbert Hoover — studied engineering and architecture at Yale University and the Colorado School of Mines. The Hoovers bought 46 acres in 1995 and built a deck, put a grill on it, and spent a few years just looking at the land to figure out what they wanted to do. They initially dreamed of a log cabin, but after Andy sat on his deck and watched slurry bombers and helicopters fighting the Buffalo Creek Fire, four miles away, the following year, he told his wife, "I don't think we're going to build a log cabin. I think we're going to build something out of stone and steel."[2]

They sited their house on a knoll to take advantage of a panorama that swept from Denver International Airport in the northeast to Pikes Peak, nearly 50 miles to the south, during the day and included, at least for a few years, hardly a light at night.

The Hoovers and their neighbors were confident that they would be warned of wildfires in the area by a "reverse 911 system" that Jefferson County dispatch would use to call homeowners when a fire threatened the neighborhood. Nonetheless, Andy and Jeanie built their house out of insulated, flame-resistant concrete, rock, and steel. They installed a 24,000-gallon water tank on the slope below the house, a 1,200-gallon cistern inside it, and another 1,000-gallon tank in a nearby shed. With his School of Mines training, Andy approached the threat of wildfire methodically. He read fire reports and bought a scanner to monitor fire calls. He cut down hundreds of trees around the property to create defensible space, and hired the sons of his neighbors, Scott and Ann Appel, to lop off low branches on the trees they left around the house, thus eliminating the ladder fuels that could turn a ground fire into a crown fire.

Historic treasures from his grandfather's presidency filled the house: Gold-inlaid flintlock dueling pistols and a hunting rifle. The president's smoking pipe. Cases of books, including a mining text from 1556 that President Hoover, who was also a mining engineer, translated from Latin into English. China that the president had brought home from Beijing shortly after the Boxer Rebellion, including an urn from the imperial palace. The porcelain alone, Andy said, was worth more than his entire house.

Andy built and repaired historic furniture, and he was restoring his grandfather's fishing gear cabinet in his elaborate basement woodshop. A nearby wine cellar held two brandies with Napoleonic seals from 1818.[3]

Eventually Andy planned to give it all to the Herbert Hoover Presidential Library and Museum in Iowa. For now he enjoyed living with his legacy.

He kept binoculars at the ready to scan for wildfires and often spotted them before any alarms went out. When the earlier Colorado State Forest Service burn on the Denver Water land jumped its containment in October 2011, the Hoovers' home on the south side of Kuehster Road looked right down on it.

Despite his diligence, Andy wasn't the first resident to report the slop-over on Monday, March 26, 2012. At 12:43 that afternoon, more than an hour before the calls to the Elk Creek and North Fork fire departments, a caller to Jefferson County dispatch reported, "I'm looking down toward Platte River Road. I think the prescribed burn down there is fired up again."

When the dispatcher asked if the resident saw gray smoke, the caller responded, "Whitish gray. It's flat, going with the wind."

The dispatcher said he would ask the fire department to look into the source of the smoke. But as the smoke and 911 calls thickened, dispatchers were less helpful.[4]

Kim Olson grew up having to practice escaping from her home on her hands and knees and memorizing meet-up points. She was just as diligent now that she had her own home and family. She often called the lookout in the fire tower at Devils Head to find out where the smoke she smelled was coming from.

"I'm the first one to call [911] whenever there's a lightning strike," she said. "I've called a lot . . . and I [wonder] if that had something to do with how I was treated that day."

March 26 was the first day of spring break, so she and her husband both decided to stay home with the kids. Doug was outside washing his car as Kim sat at the computer and telephone trying to find out what was causing the smoke she smelled.

She checked more than two dozen websites, rang her neighbors, and called the local fire department. "You don't need to call every time you smell smoke," the dispatcher told her when she called 911.

At 2:21 Sam Lucas called 911 as well. "It looks like there's a fire right at the foot of Cathedral Spires," he reported.

The dispatcher interrupted to say it was a controlled burn that flared up.

"We've got 79-mile-per-hour winds and they've got a controlled burn?" Sam responded. "Oh, wonderful. Thank you."

Thirteen minutes later Ann Appel, a normally spry 51-year-old who was housebound and frail while undergoing chemotherapy, called 911 to ask about the fire that was clearly burning somewhere nearby. "It's blowing smoke right over my house," she said with a nervous laugh.[5]

Her husband, Scott, a contractor, was away on business. Leaving the house would be difficult, not only because of her illness and treatment, but also, according to a friend of hers who called dispatch as well, because her car had a flat tire.

As with the Lucases, the dispatcher who responded to Appel's call told her the burn was nothing to worry about. Dispatchers advised nobody who called about the fire to evacuate.

At 2:31, however, three minutes prior to Ann Appel's call and 40 minutes after Chiefs McLaughlin and Rogers headed to the slop-over, dispatch asked a third area department, Inter-Canyon, to investigate the smoke reported from Kuehster Road.

Dave Brutout, a firefighter with that department, was already on it. He'd just finished lunch near Denver when he received a text from the sheriff's office alerting firefighters that the prescribed burn had escaped.

Brutout, a local rancher, had been a volunteer firefighter in the area for more than 25 years. He had worked on prescribed burns with state foresters in the Lower North Fork, and he knew many of the residents on Kuehster Road. Although the text noted that "fire spread was minimal," he suddenly had a feeling that he should head up to the neighborhood rather than go to the fire station or look for the source of the smoke. "I took a 90-degree turn east," he told me. "Just poking around to see who was home and who was at work."

He arrived around 3 p.m. and could see that the fire had moved into the valley below. It was so windy he had to hold on to his helmet. The smoke plume was headed east — straight toward him — but it wasn't yet leaning over his head.

"I tried my damnedest to make contact with the group that was on scene [at the fire]," he said. "We were surprised they weren't calling us. The Hail Mary just didn't happen."

The rest of his department was sent to head off the fire before it got to Pleasant Park. Brutout knew they had no chance of doing that. "It was utter suicide to send 10 or 20 trucks down that road," he said.

There was still no call for an evacuation, but Brutout took it upon himself to initiate one. "People were gathering, and I had them all look down into the valley at what was boiling below," he said.

By 3:20 he was advising residents he found on Arrowhead Springs Trail and then Elk Ridge Road to get out. Around 4:20 he started going door-to-door. When one woman asked when the fire would get there, he responded, "Could be two days or two hours."

Or two minutes.

On Eagle Vista Drive he advised Sam Lucas to evacuate. Sam and Moaneti loaded their pickup truck, and Sam called his son-in-law to say they were getting out. Their neighbor Eddie Schneider decided to leave after Brutout visited, and he called the Lucases to ask if they were evacuating, too. "Oh, yeah," Sam told him. "Just waiting for the 911 call."

Schneider later reported that he never received a call.

At 14141 Broadview Circle, Brutout encountered a chain across the bottom of a long driveway and heavy machinery nearby. Trees shaded the drive. Brutout turned away rather than risk getting trapped. Inside the house at the end of the drive, Ann Appel, despite her call to 911, couldn't know what was racing up the mountain toward her.

AT THE SCENE OF THE ESCAPED BURN, Chief McLaughlin assigned one of his assistants, Joe Page, as a lookout on a ridgeline. He stationed an engine along the flank of the fire. He had firefighters spray a wet line around it, but the timber dried out and reignited minutes after it was soaked. He assigned firefighters to dig a "scratch line" — a fast, improvised fire line — but the fire was too hot and fast for the firefighters to stand near it.

Communication disconnects continued to plague them, recordings of radio traffic later revealed. A supervisor trying to send an engine crew, ambulance, and brush truck held them back because he couldn't reach any leaders on the scene. "I have additional resources," he called over the radio at 3:57. "I'm holding them until we have a definite assignment."

McLaughlin warned dispatch again to prepare for evacuations. "We're working toward the trigger point for evacuating Pleasant Park and Kuehster Road," he reported.

The gully between the fire and the slope it had to climb to reach Kuehster Road was the boundary at which the fire passed out of CSFS control. As soon as the fire crossed into their jurisdiction, Rogers and McLaughlin would ask the sheriff to start evacuations.

The chiefs hoped that people in the neighborhood weren't waiting for that order, but they knew that many residents wait for authorities, or for technologies like reverse 911 calls or text messages, to tell them to get out. Homeowners tend to stay in their properties until the last minute. By then it's often too late to leave safely. They can find themselves trapped in traffic on smoke-filled roads.

At 4:41 Jefferson County dispatch reported that the fire had just leapt the gully and was making a "major run" that would necessitate evacuations. In seconds the blaze that crept down into the gully was leaping from treetop to treetop up the other side.

McLaughlin and Rogers immediately let Michalak know that they were taking over the fire. At 4:54 they called to get everyone out of Pleasant Park and Kuehster Road.

"Be advised we just requested Jeffco to initiate an evacuation," McLaughlin called over the radio. "This thing is off to the races."

"It's a running crown fire at this point," Rogers reported six minutes later.

McLaughlin looked up at the ridgeline for his lookout. He could see his portly assistant jogging as fast as he could from the flames.

"He was literally running," McLaughlin reported. "Running for his life."

The chief knew that anyone in the neighborhoods beyond his lookout would soon have to do the same. Rogers estimated the fire would reach the closest of the homes within two hours. Instead, the first homeowner would report his house burning 13 minutes later.[6]

McLaughlin ordered firefighters to give up trying to put the fire out and instead try to save the homes it was headed for. He called for five strike teams — 25 engines — to protect structures and assist with evacuations.

By the time that call went out, two other fires were threatening homes in Jefferson County, and many of the fire trucks that had started out for the Lower North Fork were diverted to the other blazes. "Instead of them showing up in 45 minutes or an hour," McLaughlin said, "resources were taking three and four hours to get here."

There were other things that weren't going as planned. Many of the reverse 911 calls notifying residents of an evacuation rang phones far from the blaze. Changed or disconnected phone numbers, difficulties matching incident maps with phones, short-term renters, residents who didn't sign up, and technological glitches gave Colorado's various reverse 911 services a 50 percent call completion rate on good days. This wasn't a good day. Wayward evacuation notices went to residents hundreds of miles away from the fire. The dispatchers, already overwhelmed, now had another flood of calls from confused people who mistakenly received notices to flee their homes. And almost nobody who actually needed to get the calls — Ann Appel and the other residents around Kuehster Road — received any evacuation notice at all.

As smoke from the Lower North Fork Fire filled the sky, an official driving through the neighborhood advised Sharon Scanlan not to worry, but she decided to round up the animals she and her husband, Tom, kept. By the time she loaded their last skittish horse into a trailer, the fire was in their pastures.

"I knew that probably within five minutes my house was going to be gone," she said.

Kristen moeller spent much of the morning and early afternoon watching the smoke from the window of the dream house she and her husband, Dave Cottrell, had spent years searching for. She never took the view of the mountains and crags for granted, even on a day like this, when the wind was so strong it upset their dogs.

Kristen was due to catch a flight to California, but rather than packing for her trip, she packed up what she could from her house. A friend came by to help. When Kristen's husband, Dave, called to inquire what was up with the fire, they could see flames in the forest below the house. Kristen broke down and handed off the phone.

"You've got 30 seconds to tell me what you want to save," their friend told Dave.

Kim olson was a little annoyed that her husband wasn't taking the situation as seriously as she was. "You don't need to panic," Doug told Kim as he washed his car. "We've seen this before."

On their deck, Kaleb, their oldest child, was cleaning the car's floor mats.

During the Hayman Fire, 10 years earlier, Kim and Doug had seen the red sky and put wet towels around their doors and windows to keep out the smoke. This didn't seem as bad, but not knowing where the fire was terrified Kim.

Kuehster Road residents hadn't been notified of the controlled burn the previous October. That fire had also escaped, she had learned. A sheriff's deputy who had noticed the fire spreading behind a locked gate had called the state foresters back to put it out. Kim couldn't remember the CSFS ever having notified the residents of Kuehster Road of their nearby burns — or when they escaped. "Never, ever," Kim said.

Other residents drove through the neighborhood to try to get a fix on where the fire was and where it was going. Kim got out a topo map and looked at where the earlier escaped fire had burned. The wind was blowing eastward, and her house was directly east of the previous blaze. "That's when I started throwing stuff in the car," she told me.

She had her daughter, Rhoanabella, age nine, videotape the house's contents and called Doug to the computer, where Pinecam.com was putting up all the scanner calls related to the fire. He called their neighbors Dave and Carol Massa, who had called 911 about the burn the previous Thursday and reported more smoke coming from the Lower North Fork over the weekend. Doug told his neighbors that he didn't think there was anything to worry about.

Kim, however, was adamant. "I'm leaving," she told him. "This is the big one."

As they led the kids outside, with Doug carrying their youngest, Quillan, age four, the sky had turned from gray to a dark black and red swirl. "In 15 or 20 minutes it went from something we had seen before to a red hurricane," Kim said.

A freight-train roar came out of the woods around their home. Big chunks of debris from houses, along with paper and ash, started to rain down on their yard as they ran to their cars. "That was probably the Appels' house," Kim realized, unaware that Ann Appel had also called 911 and been advised that there was nothing to worry about.

Kim could see fire at the end of her driveway. "It's too late," Kim thought. "We're not going to make it."

Doug threw the keys to their Jeep to Kim and loaded the kids and their dog into his car. Kim led the way in the Jeep up their drive and out to the road. "This is stupid," she thought. "We should all be in the same car."

Doug's car was filled with the panicked panting of his children. He handed his smartphone to Kaleb to record a video.

"Daddy," Quillan gasped, as sparks and embers flew past the car and bounced off the windshield.

"We'll make it," he told them. "We're going to be fine."

"Where's Mom? What's she stopping for?" his son cried.

The only way out of Kuehster Road was to turn right at the top of their driveway. But the dead end to the left was, for the moment, clear of fire and smoke. Other escape routes for residents of Pleasant Park involved unmaintained tracks behind gates that were often locked. Kim started to turn away from the smoke onto the dead end, but another car raced out from it. It was Eddie Schneider, who had evacuated after his last conversation with the Lucases. Later, she would nickname him "Lucky Eddie," but at the moment Kim had no idea who was in the car in front of her.

"We should try to make it through," she thought. She turned right and pulled in behind the other car.

Ahead, the road vanished into a point of orange that looked like the light at the end of a tunnel but offered anything but hope. Doug followed, pointing out the red glow in the woods just below them on the right side of the road. "It's down there," he said as his son pointed the phone. "It's down there now."

The crowns of trees just to the left of the road burst into flames. The children gasped as the fire closed in on them. "There it is," Doug shouted, "right here."

From her father the firefighter, Kim had learned that an exploding wildfire can suck so much oxygen from the air that cars stall. The heat can melt their tires. She sped up into the blackness.

"We're out, we're out, we're out," Doug shouted to his kids as they passed beyond the wall of smoke.

Kim and Doug stopped beside a sunny pasture just outside the forest, embracing each other and their children as they looked back at the black smoke mushrooming over the road they had just driven on. A motorcycle vanished into the smoke, then raced back out seconds later.

Doug turned to his wife. "I think I just killed Dave and Carol," he said, recalling that not a half hour earlier he had nonchalantly told his neighbors that the smoke wasn't anything to worry about.

"We have to go," Kim said as the smoke rolled toward them.

When Kim realized that her husband had had her son shoot a video of their escape, she was furious. "How could he detach himself from the crisis like that?" she later fumed.

But within days, as the video appeared on screens around the world, she realized how important the document Kaleb had created was.[7]

DAVE BRUTOUT, GOING DOOR-TO-DOOR to warn residents, saw the fire roll onto the south end of Kuehster Road shortly after 4:45 — about four minutes, he guessed, after he'd finished talking with Sam and Moaneti Lucas. "It was just folding over on top of itself," he said. "What it was curling over was already bursting into flame."

It crashed up the road from the west in four or five waves, each one rolling farther north into the neighborhood. The flames leapt 100 feet into the sky and ran 200 feet per minute. "It took out a 10-acre [swath] at the end of Kuehster Road, and then it took [out] another 10-acre swath, and then another," Brutout said. "The first one hit me in the ass as I was leaving."

The wind blew off his helmet, chin strap, and goggles. A firebrand struck him in the eye.

"Kids were running down the road with their stuff in plastic bags," he told me. "Their bags were melting. One resident drove off the road and off the embankment and had to climb into a fire truck."

Some people were headed back into the neighborhood, so Brutout blocked the road.

"Residents in town saw the smoke and drove around roadblocks and through creeks and through fences to get to their homes," he said. "A deputy drove . . . at 40 miles per hour past my roadblock into a smoked-out road."

Jefferson County deputies David Bruening and Jason Hertel helped a homeowner on North Trail Court load his horses into a trailer. Then the man announced, "I'm not leaving. I've built this all by myself."[8]

The deputies couldn't make him leave, but they weren't staying. Around 6:20 they and two other sheriff's deputies lined up their four cars in a caravan and began driving out Kuehster Road toward Pleasant Park. Visibility was near zero, with ash, smoke, and flames hiding the road. Deputy Bruening navigated by GPS in his car but veered off the pavement. He pinballed in and out of a ditch, then hit a small tree. When he tried to back away from it, the car got stuck on some large rocks. Flames surrounded him.[9]

The other deputies turned on all their lights to guide Bruening to their vehicles, but in the dense smoke he still couldn't see them. Deputy Jerry Chrachol pulled behind the stuck patrol car and radioed Bruening, who ran to the other vehicle and dove in. The cars retreated to an open pasture and waited for a fire truck to lead them out.[10]

ANDY HOOVER WAS HOME ALONE when he heard the fire chiefs recommend an evacuation over his police scanner. He called his wife to tell her what was happening and repeatedly called his friends the Lucases, across the street, confirming that they planned to evacuate.

A few minutes after they last spoke with Andy, Sam and Moaneti were ready to pull out, but Sam ran back into the garage, where he had a fire suppression system that could cover his house with foam. His wife waited outside.

Meanwhile, Andy moved furniture and valuables away from windows, then, although he was confident his house could endure the flames, loaded up his truck to escape. He saw the sky flashing black and orange. Sparks and firebrands filled the air. It was safer inside the house, he decided. "This might be my last day," he thought.

He hunkered down in a hall as the firestorm raced over his house, which heated up like an oven. Window glass cracked, shattered, or blew out. He heard a propane tank explode.

The tempest moved fast. Andy saw flames on his decks. He went out and doused one, but the fire on the other, outside Jeanie's greenhouse, grew quickly. With the power out, Andy lost water pressure before he could extinguish it. When he went back in, the house was filling with smoke. Something was burning inside. He got on his hands and knees and made his way to the garage. Andy started his truck, but the garage door didn't respond when he pushed the opener. He pulled at the door but couldn't budge it, so he got back in his truck, floored the accelerator, and smashed out of the garage. "They don't build them like they used to," he joked afterward.

Outside, trees were charred, but the flames were subsiding. A mountainside covered with spot fires looked like the Milky Way. Andy stopped his truck and looked back at his house. Slowly the windows lit up orange and flickered with flames, like a jack-o'-lantern's wide eyes and grimacing mouth. The house built to endure a fire on the outside burned from the inside. He saw the president's library burning. Herbert Hoover's china, guns, wine, and furniture would soon be rubble.

Andy called a 911 dispatcher to report that his home was burning but declined the dispatcher's advice to keep driving out of the neighborhood. "I think I'm going to hang out here until I know a better idea," he said. He was safely in the black — parked on pavement surrounded by charred ground. "It seems like a dumb idea to move."

Andy turned up the air-conditioning in his truck. Around 7 p.m. he stepped from the cab to shoot a video of the flames exploding through the roof and windows of his house. Across the street, the Lucases' home was a blackened heap with a burnt truck outside it. In the driveway Moaneti still waited for her husband to return from the collapsed, smoldering garage.

12

RED ZONES

DAVE BRUTOUT, THE FIREFIGHTER who had warned the Lucases and many other homeowners to evacuate, met up with his captain near the radio towers along Kuehster Road. "There are people that did not make it out of there," Brutout told him.

With the winds and the temperature dropping, the flames were starting to calm, but there was smoke in the air, fire in the trees, and rubble on the ground. Brutout walked back into the neighborhood in front of two fire trucks. In a long driveway he moved fallen limbs so they could get an engine to the burning house.

"When I moved the fourth one, it wasn't a log. That was the resident I'd talked to a few minutes before," he said of finding the Lucases. "To look at my friends [lying in] the driveway . . ."

FROM MY HOME IN BOULDER, 45 miles to the north, the smoke of the Lower North Fork Fire rose like a thunderhead out of the Front Range of the Colorado Rockies.

At Bill McLaughlin's firehouse, television satellite trucks and reporters crowded the parking lot, while inside firefighters and sheriff's deputies scrambled to put together their incident command post. I stood in the station's garage, empty of trucks, as leaders of the incident clustered in a corner away from the chaos.

Three people were dead. The Lucases had already been found, but Ann Appel's body, buried in the rubble of her home, wouldn't be discovered for days.

Some evacuees moved in with family and friends nearby, but most gathered at a motel a few miles from Pleasant Park. They frantically tried to account for their neighbors — knocking on the doors of one another's rooms, calling one another's cellphones, and conferring in the lobby. Some were unable to speak and others incapable of stopping. Most had no idea whether their own homes were burning. Kim Olson and Doug Gulick's house survived, although their property was badly burned. Tom and Sharon Scanlan, and Kristen Moeller and Dave Cottrell, would lose their homes.

Whether they were officials wearing a uniform or evacuees wearing the only clothes they owned after the fire consumed all their belongings, they had one question in common.

"How did a fire set to protect us turn into our destroyer?"

WHILE OFFICIALS AND RESIDENTS POINTED to the overgrown forest, the drought, confused communications, failing emergency notification systems, and the very wisdom of lighting a forest heavy with fuel and parched with drought on fire, another part of the answer could be found simply by looking at a map and the local census. It wasn't just trees overcrowding the forest.

Laura Frank was an investigative reporter at the *Rocky Mountain News* until that newspaper's sad closing in 2009. The following year she and I were journalism fellows together at the University of Colorado. We spoke often about the state's rapidly expanding wildland-urban interface, or WUI — the term firefighters use to describe the area where communities abut fire-prone landscapes. During her fellowship Laura founded I-News, an investigative journalism operation, then hired Burt Hubbard, another veteran investigative reporter from the *Rocky Mountain News* and *Denver Post*. Burt took maps that the Colorado State Forest Service had made over the previous decade showing the state's "Red Zones" — its most flammable forests — and overlaid them with maps of Colorado's previous three censuses.

When I saw his maps, just a few weeks before the Lower North Fork Fire, I was dumbfounded. On Hubbard's computer screen the expanding population rushed into Colorado's most fire-prone forests like an invading army.

Between 2000 and 2010, more than 100,000 people had moved into the state's Red Zones. When we looked back another 10 years, to the 1990 census, that number rose to a quarter million new residents. By 2012, 1.1 million Colorado residents and half a million homes — that's one in four homes in the state, and one in five residents — risked burning in a wildfire.

In Jefferson County, where the Lower North Fork Fire burned, 28 percent of the population lived in a Red Zone, a 24 percent increase over 10 years earlier. In other Colorado counties like Summit, Teller, and Pitkin — home to Aspen — more than 90 percent of the residents lived in Red Zones.

Across the West, the story was the same. A 2015 report by CoreLogic, a company that analyzes data for the global real estate and home insurance industries, identified 1.1 million homes in 13 western states as highly vulnerable to wildfire. Rebuilding those homes would cost $296 billion.[1] CoreLogic counted only homes in specific zip codes that were built with flammable materials and located close to vegetation that could put embers on their roofs. A 2013 study by the U.S. Forest Service's Rocky Mountain Research Station noted that nearly a third of all U.S. homes were in the wildland-urban interface.[2]

So it should come as no surprise that four times more homes burned in U.S. wildfires in the 2010s than in the 1990s. In 2008 the International Code Council reported that, on average, 932 structures burned in wildfires each year during the 1990s, but in the first eight years of the twenty-first century, 2,726 structures burned annually.[3] From 2011 to 2016, according to the National Interagency Coordination Center's annual wildland fire reports, an average of 3,754 structures burned each year — quadruple the average of the 1990s.

As frightening as those numbers are, they're likely just the beginning.

In 2012, according to Headwaters Economics, a think tank in Bozeman, Montana, the cost of protecting homes in Montana from wildfires was $28 million. At the current rate of development, that figure would increase to $40 million by 2025. Add a one-degree increase in summertime temperatures — which researchers predict would lead to a 125 percent increase in the number of acres burned in wildfires in the state and a doubling of the cost of protecting homes from those fires — and that one state would need to spend $84 million a year defending homes from forest fires.[4]

In the 1990s, according to the Headwaters research, the average cost of wildfires to the federal government was less than $1 billion a year. In the decade leading up to the Lower North Fork Fire, that average increased to more than $3 billion.[5]

Then there's nearly $2 billion spent annually by state governments, and similar costs borne by local governments and fire protection districts, like the Elk Creek, North Fork, and Inter-Canyon fire departments that responded to the Lower North Fork Fire.

Yet 84 percent of the private land in the WUI of the western United States remains undeveloped. Ray Rasker, executive director of Headwaters, estimates that if just half of that land were developed, the costs of fighting wildfires for the federal government alone would run as high as $4.3 billion a year. That's nearly 80 percent of the Forest Service's average annual budget.

Headwaters' studies show that 30 percent of all the money spent on wildfires is associated with keeping homes from burning. Currently, the U.S. Forest Service estimates that protecting homes accounts for somewhere between 50 and 95 percent of its funds for fighting wildfires.

While Laura, Burt, and I studied the Red Zones, Tania Schoennagel, a researcher at the Institute of Arctic and Alpine Research, was looking at how areas zoned for future development are at even greater wildfire risk.

Earlier housing developments in Colorado's Red Zones were generally at lower elevations, had more widely spaced trees, and were located in less steep terrain — those parklike ponderosa pine woodlands that Mike Battaglia showed me. The fires in many of those forests, if excess fire suppression didn't let them grow unnaturally thick with vegetation, tend to burn along the ground with low intensity. In most of those forests, if they were restored to their historic density and health, homeowners would likely confront more manageable, less intense fires.

But new developments in the Red Zones are increasingly at elevations above 8,000 feet, in forests often dominated by lodgepole pines that naturally burn in large, intense crown fires that destroy entire stands of trees. Other, "mixed-conifer" forests, which grow above about

7,500 feet, depend less on fire than lodgepole forests, but still burn more intensely than most ponderosa pine forests. Steep slopes and chimney-like canyons in the high country can magnify the intensity of the fires in lodgepole and mixed-conifer woodlands.

"As you move up in elevation you get . . . very dense forests — those lodgepole pine forests," Schoennagel told me when I visited her. "Characteristic of those are these high-severity fires that happen very infrequently and burn through the treetops. They're very difficult to fight."

Topographic maps show that all of the homes destroyed in the Lower North Fork Fire were above 8,000 feet. But the fact that more trees burn in a forest doesn't mean that more homes have to.

Jack Cohen, of the Missoula Fire Sciences Laboratory in Montana, has walked through neighborhoods across the country after wildfires reduced them to rubble, and crawled through countless incinerated homes on his hands and knees trying to track down the culprit in their demise. He's seen hundreds, like Andy and Jeanie Hoover's, that ignited after the fire passed, sometimes hours later. Often green, unburnt trees surrounded the charred houses.

Cohen came of age in the 1950s and '60s, and his obsession with what he's come to call the "home ignition zone" grew out of his own playing with matches as a child. "I really didn't know I was a prescribed fire–lighting boss when I was 12 years old, but that's basically what I did," he told an audience at a fire conference in Denver a few months before the prescribed burn exploded onto the homes of Kuehster Road.

Cohen grew up in Southern California's chaparral, where both fuels to burn and the supplies to ignite them were easy to find. "In the fifties," he said, "everybody smoked, so there were matches everywhere."

He lit a teddy bear cholla, or jumping cactus, near the outbuildings on his family's property. "I wasn't totally stupid," he said. "I had a hose across the fence and turned on. But by the time I got the hose, [the cactus] was gone."

The fire burned a swath of the family's land but didn't destroy any of their buildings. Years later, as a college student studying wildfire, he watched a film of the 1961 Bel Air–Brentwood Fire.

"I had this impression of wildfire spreading to and igniting the [houses]," he said. "The video showed flames doing that, but also igniting houses one to two blocks away."

In fact, in Brentwood houses more than a mile away from the fire ignited.

"It's not necessarily the big flames [that are the most dangerous]," he told me. "That's the thing that captures our attention. It's clearly the thing that would threaten us the greatest as people. But . . . one of the products of those extreme fire behavior situations are firebrands. People tend to ignore the firebrands because [they're] not mortally threatening."

While 100-foot flames in the trees may pass by a home or neighborhood, a few needles in a gutter, grass under a deck, or a broom leaning on a porch can catch a tiny ember that will ignite a house. Or a puddle of flame, just inches high, can work its way to the wood siding.

"Instead of thinking about it in terms of the big flames causing the ignitions, let's think about it in terms of the massive amounts of firebrands that are lofted into the residential area," he said, "and the houses themselves becoming the primary fuel source."

Firebrands can fly more than a mile, so keeping the forest far enough from homes so that no embers could land on them would require massive clear-cuts. Cohen doesn't think that is necessary. Rather than cutting down all the trees, according to Cohen's research, diligently picking up tree needles, keeping woodpiles and propane tanks far away from structures, and installing metal roofs are more effective methods of keeping homes from burning. "We can . . . have extreme wildfire behavior and still not necessarily have . . . a residential fire disaster," he told me.

Yet in the Lower North Fork Fire, many of the owners of homes that burned, like the Hoovers, had done all of those things.

When Tom and Sharon Scanlan began building their home in the Kuehster Road community six years before the Lower North Fork Fire, they could see mountainsides scarred by 2002's Hayman Fire. "We looked at it every single day," Tom told me as we walked through his charred home and woods after the Lower North Fork Fire.

Tom, a retired U.S. Air Force general who oversaw rocket launches, knows a little bit about combustion. Those charred mountainsides inspired him to build his mountain home out of fire-resistant, insulated concrete; to bury the propane tank far from the house; and to cut scores of trees to create defensible spaces around the house and barn.

"We often had guys come up and talk to us from the state tree forestry," he said. "We understood the risks, and we . . . did everything . . . more than was recommended. I was cutting down trees every year."

Even so, the Lower North Fork Fire left the Scanlans' home a charred ruin.

"There was no mitigation that could possibly be done to stop what happened here," he said.

COLORADO, LIKE THE WEST as a whole, saw a marked increase in the number of wildfires during the 2000s. State records show that 217,960 acres burned from 1990 through 1999. In the following decade, from 2000 through 2009, 889,343 acres burned — a fourfold increase.[6]

What caused this increase? The warming and drying climate extended the fire season. Excess suppression of fires and resistance to prescribed burning and thinning projects left some forests with dangerously heavy fuel loads. But what turned the subsequent fires into disasters was the booming population in the forest.

THE UNITED STATES HAS DEVELOPED a vast collection of cutting-edge technologies that can be used to look at how the weather, climate, vegetation, topography, and human development drive wildfires around the planet, as well as to manage the response to them. At the National Interagency Fire Center in Boise, Idaho, I stood in a room the size of a small grocery store amid an array of computer screens and large television monitors overseeing the nation's fight against wildfire. In its ability to monitor hundreds of fires at a time (as well as to respond to everything from hurricanes to terrorist attacks), to oversee as many as 30,000 firefighters, and to manage everything from shovels to satellites, the command center is often compared to a war room at the Pentagon.

After observing some teams monitoring the weather and climate, and others managing hundreds of fire crews, dozens of aircraft, and thousands of other pieces of equipment sent to wildfires around the country, I walked through warehouses of portable weather stations, communications equipment, and airborne fire surveillance and suppression technologies. In Missoula, 350 miles away, I watched scientists create fire whirls in a test tube the size of a small silo. But as impressive as those resources were, in the weeks that I studied the maps of population expanding into Colorado's Red Zones, I also learned that the flashy science and cutting-edge technologies available to study and respond to wildfires don't always translate into advantages in fighting them.

I saw one of the more colorful firefighting technologies in a baby pool that sat atop a table in the Grand Ballroom of the Boulder Marriott. The blue tub decorated with cartoon octopuses and sea horses was half-filled with sand. From a pole above it, a digital projector and camera pointed straight down into the pool.

Rodrigo Moraga turned on the projector, spraying splotches of colored light onto the sand. The words "Wildfire" and "Black" appeared at the edge of the projection, along with backward and forward arrows, like those that control a tape player.

Half a dozen firefighters pressed in shoulder to shoulder around the pool as the ballroom lights dimmed. The training conference for emergency responders was planned before wildfires ravaged the state, but it was nonetheless an indication of a community awakening to how flammable it was. To Moraga, Boulder County's residents couldn't wake up fast enough. He was the first firefighter to respond to the Fourmile Canyon Fire, which broke out on Labor Day 2010 and was the first of four fires in four years that would set records for destruction and expense in Colorado. To Moraga, the smoke had hardly cleared before the Fourmile Canyon Fire was forgotten. But he wasn't likely to forget it anytime soon.

He had hoped for a better turnout at his demonstration of the Simtable, a wildfire modeling tool. Yet the audience he managed to gather responded to the light show with the enthusiasm of high school boys playing a new video game, leaning in with murmurs and gasps.

He stabbed his hands into the glowing sand to model the Boulder Range according to an elevation profile projected by the simulation, then waved a laser pointer like a magic wand to raise a vision of the mountains above Boulder like a hologram. Forests glowed in different shades of green, grasslands in yellow. Some buttons and dials made of light controlled the weather in the simulation—wind speed, relative humidity, and temperature. Others assigned bulldozers, different types of aircraft, fire engines, and crews of firefighters. With his pointer he drew in fire lines, fuel breaks, and, finally, fire.

He used the laser to hit a Play button to start a time lapse, and the colored lines, dots, and splotches began to move. The fire grew in response to the weather, fuels, and firefighting efforts programmed into the simulation. Lines growing around it reflected the training and fitness of the crews he had chosen to fight the blaze and the terrain where they were working.

Moraga pointed his laser at the compass rose indicating wind speed and direction, spun its arrow, and pulled it longer, thus turning the gusts and increasing their force. The fire overran the line of firefighters.

The scene looked like the work of a sorcerer, and Moraga fit the part. He's five feet six and wiry, with a shaved head and a soul patch that can make him look impish one moment and sinister the next. Rather than a fire department T-shirt, he wore the type of short-sleeved plaid shirt popular with the town's rock climbers and mountain bikers.

He called up a fire that burned in the mountains above Santa Fe. The city's grid projected on the sand reminded me of the nighttime view of Boulder from my porch. White dots representing cars marched like ants along the grid of roadways in a simulated evacuation from the spreading fire. Highways turned into gorged white lines as they backed up with gridlock. Eventually the red blotch of fire seeped over the thick cords of traffic.

"Those people are toast," Moraga said.

"You just destroyed a town," one of the firefighters watching the simulation said. "How do you feel about that?"

Moraga looked up with a wicked grin. "Pretty good," he answered.

When I asked him if he'd run a simulation of the Fourmile Canyon Fire, his smile faded. "It wouldn't have made a difference," he said. "Except that I would have put a lot of retardant on my house."

He handed the pointer to another firefighter, who set the program to read the laser flicking the sand as lightning strikes above Santa Fe. Other firefighters just held their Bic lighters to the sand to start new simulated blazes.

Moraga was born in Chile but grew up in Montclair, New Jersey. He came to Boulder to work as a ranger, with dreams of getting paid to backpack. Instead, he found himself fighting fires. He was on the first controlled fire the county set outside Boulder, then became its prescribed burn manager and an ignition specialist. After training as a fire behavior analyst, his first assignment was on the Hayman Fire (2002) — the largest fire in Colorado history. "Fire was my friend," he said.

But he learned he preferred modeling it on a computer to fighting it with a Pulaski. "I was a terrible firefighter," he said. "Digging line is not for me. But I can understand it. I knew what it was going to do. I was lucky to find an adventure that allowed me to use my brain and be on the fire, too."

A company he cofounded, the Anchor Point Group, specializes in reducing the threat of wildfires to communities. The fact that he lived in a house along Fourmile Canyon was just another layer of expertise.

When the Fourmile Canyon Fire broke out on Labor Day 2010, Moraga was the first firefighter on the scene. "I called the chief. 'Just send me an engine and a five-person hand crew,'" he said. "I didn't think it was that bad."

At a table across the ballroom from the high-tech simulation, Moraga drew the Fourmile Canyon Fire on a piece of hotel stationery. He marked the bonfire lit by an elderly volunteer firefighter that started the blaze, drew a long arrow to show the downburst of wind he believes blew it up, and then scrawled a blotch shaped like a holly leaf to show the spot fire it ignited.

The steep canyons above Boulder are perfect chimneys, and the winds on that Labor Day gusted hard. They squeezed the fire like springs. The flames bounced back up the canyons to the north, east, and west.

With flicks of his wrist, Moraga scratched lines to the edge of the paper from the blotch of fire. The line across Fourmile Canyon Drive to the gorge's north-facing slopes was straighter and darker than the rest.

Northerly aspects above Boulder are generally wetter, which, counterintuitively, often makes them more likely to burn. Fire-prone Douglas firs flourish on the cold, moist slopes, and the wetter soils support more vegetation to fuel a fire. After years of fire suppression, some of the ponderosa pine forests in the canyons had more than 10 times the number of trees they would naturally carry.

Property owners with no view of their nearest neighbors made for another unhealthy crowd. According to a 2013 report from Headwaters Economics, Boulder County has developed more than half of its wildland-urban interface — the highest proportion in Colorado. The 5,409 homes that Headwaters identified in Boulder County's WUI increased the fire risk just as much as the unhealthy forests.[7]

The land is a jigsaw puzzle of old mining claims, many now occupied by private homes, cut into public forests. Steep, zigzagging roads — some of them bulldozed without permits by the residents themselves — are difficult for fire trucks to navigate.

After telling a local reporter that the fire was "no big deal," Moraga stopped by his own home, a 2,300-square-foot ranch set on five acres of pines and firs.

"I wasn't very concerned about my house," he said. The blaze was above it, and "fires back down slow."

He'd cut down the trees to create defensible space, sited the propane tank far from the building, and kept pine needles and flammable furniture off his deck. None of that made a difference.

"When I showed up, the hillside beside my house was on fire. But I couldn't get an engine," he told me. He was in charge of assigning resources to fight the fire and had already sent all the available engines elsewhere.

"We could have saved any one of the houses that burned in the Fourmile," he said. "But when you've got all these houses threatened in different canyons, in absolutely terrible places as far as fire is concerned . . .

"I could have saved my own house with just 50 gallons of water," he said. Instead, he watched the fire creep down the hill and onto his porch.

He stayed on the fire for two days. "Which is a day longer than I should have," he said. "Considering that by the second day, my wife and kid were homeless."

Eleven firefighters lost their homes in the fire, which also destroyed the Fourmile Fire Department's firehouse, Moraga's station.

Even a fire truck couldn't save some homes. Tom Neuer, a former

California firefighter who lived about a quarter mile from where the fire started, had an 8,000-gallon underground water tank and a fire truck retired from the Fourmile Fire Department that he had purchased to protect his property. He and his new wife spent the Labor Day weekend clearing trees and brush from around their house and driveway, part of the fire hazard mitigations Tom had been doing for a decade. They were gathering up the wood they had cut down when they smelled smoke.

The crown fire came over a nearby ridgeline with 100-foot flames that raced through the treetops with a roar his wife mistook for a retardant bomber. "It didn't even touch the ground," Tom told me.

He turned on the fire truck's pump but abandoned it when he realized he had no chance of stopping the flames. "I left the engine running, and I left the hose running," he told me. "The fire boiled over the house like nothing I've ever seen."

He raced away in his van, bouncing into the air down switchbacks through walls of flame that rolled across the windshield like water and melted the side mirrors. Tom cursed loudly, recalling a fire truck driver he knew who died in a California wildfire. He survived, but his house did not.

The blaze burned 168 homes — multimillion-dollar mansions, single-wide trailers, and shacks built in the 1800s. More than $80 million worth of property burned. The National Interagency Coordination Center, the wildfire war room I had walked through in Boise, sent upwards of 1,000 people from 192 agencies and 20 states to join the fight.

Moraga handed me his sketch of the fire that destroyed his home and glanced back at the firefighters playing in the sandbox with the computer's fire simulation.

"Fire big, man small," he said. "We're done living in the mountains."

PART IV

TURNING UP THE HEAT

13

PLAYING WITH FIRE

Jamestown, Colorado — March 31, 2012

ON SUNNY SUMMER DAYS, Lefthand Canyon Drive, half a mile north of where the Fourmile Canyon Fire burned, fills with an average of 750 bicycles climbing to the hippie hamlets of Jamestown and Ward.[1] Fishermen stand in Left Hand Creek, while motorcycles and ATVs join hikers at trailheads. But to many people who live in the canyon, the sportsmen who stand out the most aren't the crowds visible on and along the road, but the ones they hear out in the woods. Gunshots on the nearby Forest Service land can make both the riders and residents nervous.

The week of the Lower North Fork Fire I biked up the canyon to find half a dozen firefighters mopping up a charred quarter acre at a turnout just west of Jamestown. It was one of two wildfires that broke out in the canyon that weekend. When I asked how the blaze had started, one of the firefighters pointed out the tattered targets hanging from some of the trees. Nearby I saw spent bullet casings on the ground.

The fires left residents of the canyon on edge about the risk of ignitions from gunshots in the woods. "It's just potentially a deadly combination," Mike Matzuk, a resident of Castle Gulch, told the *Boulder Daily Camera*, "... a scary combination for residents."

A week later yellow tape blocked trailheads in the Roosevelt National Forest. Campfires still start most of the wildfires in the woods, but bullets and hot motorcycle engines ignite plenty, the rangers enforcing the ban told me.

In the days that followed, however, the concerns of the Forest Service were more than matched by the anger of shooters and bikers who were blocked from their forest playgrounds. Some barked about their Second Amendment rights or denied that bullets could start fires. But even sports not usually associated with sparks were igniting blazes.

In 2010 a golfer chipping his ball out of the rough at the Shady Canyon Golf Club in Orange County, California, hit a rock with his club, sparking a 12-acre fire that took helicopters and 150 firefighters nearly seven hours to extinguish.[2] A year later a golfer at another Orange County course looked up from his chip shot to find himself surrounded by flames on a grassy hillside. Neither stomps from his feet nor ice and beverages from his cooler could snuff the fire, but a turn in the wind contained it against a cart path.[3] In 2014 the Poinsettia Fire ignited on the seventh hole of the Omni La Costa Resort and Spa and went on to destroy five homes, 18 apartments, and a commercial building in San Diego County.[4] One person was found dead in a transient camp burned over by the fire. Investigators never determined a definitive cause but identified a spark from a golf club as the likely culprit.

In 2012 shooters started far more wildfires than golfers. On the first day of summer in Saratoga Springs, Utah, more than 9,000 people were forced to evacuate their homes by the Dump Fire, which ignited when two riflemen were shooting in a landfill. The blaze threatened to overrun an explosives manufacturer.[5] Sheriff James Tracy also worried that it would take down the area's power grid (scores of utility poles burned). It was the 20th fire ignited by target shooting in Utah so far that year. Although the governor asked target shooters to refrain from firing in forests, a recently passed bill deprived Utah's sheriffs of the authority to restrict gun use. Firearm enthusiasts, among the most diffuse and control-resistant recreational groups, continued to blast hot lead into paper-dry woodlands.

In the Dump Fire investigators eventually implicated a product that had become increasingly popular among marksmen — exploding targets, which were igniting woodlands across the country.[6] These targets make it easier to see if a distant mark is hit, but their popularity is largely due to the entertaining flash and bang. "They tie a couple of them together and soak them in gasoline," one forest firefighting friend told me.

The Springer Fire in Park County, Colorado, which burned more than 1,000 acres in 2012, was one of three fires started by exploding targets documented in the state and 16 noted by the U.S. Forest Service in 2012 and 2013. The website Wildfire Today counted 27 fires ignited by exploding targets in those years. One killed a marksman with shrapnel, another blew the hand off a man who was mixing the explosives, and a third killed a shooter and left a 10-foot-wide crater in the Oregon woods. In Pennsylvania an exploding target seriously injured two game commissioners who were fighting a fire at a gun range.[7]

In August 2013 the Forest Service outlawed exploding targets in national forests in the Rocky Mountains. They were already banned in Lefthand Canyon when I bicycled up to the firefighters there. But many shooters use the targets on private land, and the fires they start pay no heed to forest boundaries.

As U.S. forest managers struggled to reduce that risk, another, more family-friendly form of entertainment — paper sky lanterns, which rely on a small open flame to ascend into the sky — started carrying fire even farther than the exploding targets, igniting forests and structures around the world.[8]

14

NUCLEAR FRYING PAN

Los Alamos, New Mexico — June 3, 2014

ON ANOTHER BIKE RIDE I learned about more worrisome weapons in the woods. From downtown Los Alamos, New Mexico, I biked west on Trinity Drive past the Los Alamos National Laboratory, where the atomic bomb was developed during World War II. The lab sits between the Rio Grande Gorge and the Jemez Mountains. The Pajarito Plateau rolls away from the forested peaks, its mesas spreading like tentacles toward the Rio Grande.

The lab's 43-square-mile reservation is filled with ponderosa pines, some scattered among the cliffs and gullies, some in dense stands, and some just burnt, shattered trunks. Signs on the fence around the lab warn "No Trespassing, Explosives." I wondered how fire or firefighters could interact with a landscape filled with bombs, atomic or otherwise.

In the Jemez Mountains above Los Alamos I could see remnants of 2011's Las Conchas Fire, which threatened both the town and the laboratory. When I turned back east, I descended past scars of the Cerro Grande Fire, which burned in 2000 from Bandelier National Monument, the site of ancient Native American cliff dwellings and ruins, onto the lab's land. The two blazes highlight a dangerous dilemma in the forests around Los Alamos.

For more than half a century after the founding of the nuclear lab, firefighters attacked every natural fire in the surrounding forests. Heavy grazing removed the grasses that carry the low-intensity ground fires that thin the understory. In the past, fire normally burned the area about every 10 years, but after decades without it some forests had 10 times the number of trees they had historically.

But timber wasn't the only thing making the landscape explosive. According to an analysis of the Department of Energy's "1996 Baseline Environmental Management Report" by Concerned Citizens for Nuclear Safety and the Nuclear Policy Project, the lab then had more than 2,120 "potential release sites" that included radioactive and hazardous waste disposal and spill areas contaminated with everything from radionuclides, high explosives, and heavy metals, to sewage and chemical spills. A contaminated reactor sat on the bottom of Los Alamos Canyon. A vault in Pajarito Canyon secured plutonium and highly enriched uranium. Water Canyon held significant residues of high explosives from a firing site. Shallow shafts in one "material disposal area" held some 40 kilograms of radioactive materials, including plutonium, from test explosions.[1]

Following World War II, as the fuel loads — both timber and nuclear materials — grew, so did the volatility of nearby fires. A 1954 blaze forced the evacuation of the town. In 1977 a firefighter died of a heart attack while running from the exploding La Mesa Fire. In 1996 a 16,500-acre fire, the largest in state history to that date, forced 48 firefighters in Bandelier into their fire shelters — one of the largest such deployments on record. Only a shift in the wind kept the fire from burning the laboratory. Two years later, in 1998, firefighters battling the Oso Fire, an arson about eight miles north of Los Alamos, said the blaze burned so intensely that the ground shook.[2]

The Jemez Mountains wrap around three extinct volcanoes. The largest is Valles Caldera, a 13-mile-wide volcanic bowl filled with lush forests surrounding a valley bottom of grasslands dotted with hot springs, volcanic domes, and steaming fumaroles. From the thick, high-altitude spruce and fir forests that cover the mountains' northern and western peaks, to the ponderosa pine forests sprawling among the red rock, to the piñon pines, junipers, and grasses in the canyon bottoms, the landscape encompasses as diverse a variety of vegetation and geology as can be found in such a small swath anywhere in the West.

Cerro Grande, a 10,207-foot decapitated volcano that is the high point of Bandelier National Monument, looks out on the Valles Caldera National Preserve, the laboratory, and the town. Before sunrise on July

16, 1945, workers from Los Alamos hiked up the mountain to see a flash 200 miles to the south — the first atomic bomb test. Fifty-five years later the massive cloud rising from the exploding forests of Cerro Grande could be seen from the site of the test.[3]

On the evening of Thursday, May 4, 2000, 19 firefighters from the National Park Service and the Bureau of Indian Affairs, which was assisting the NPS, climbed the peak to burn a blackline, the first stage of a three-phase prescribed burn planned to continue through the following year to prevent a wildfire that might threaten the lab. The blackline burn was ornery almost from the moment they lit it. By the next morning it had overwhelmed the handful of firefighters controlling it, and fire slopped over the line. Hotshots, air tankers, and helicopters came to help. Backburns and fire lines corralled the blaze briefly, but heavy winds drove the fire hard. A second set of burns intended to help contain the initial slop-over instead leapt into the treetops and ran fast through Frijoles, Water, and Los Alamos Canyons.

The laboratory notified the Pentagon. Some 75 New Mexico fire departments raced to the city. Nonetheless, the following Wednesday, May 10, a vicious wind blasted the fire into the community, forcing the evacuation of 13,000 people, some of whom had been so confident the government wouldn't let Los Alamos burn that they'd sat on their porches watching the blaze. A few workers at the lab weren't so confident and left not only Los Alamos but New Mexico, fearing the fire could set off conventional explosives and release dangerous radiation. As the town's residents evacuated, the roads were clogged by gridlock.

Some 7,500 acres — about one-quarter of the lab's area — burned, along with 112 minor laboratory structures, such as office trailers and sheds. According to the laboratory, none of the buildings that burned contained nuclear materials or high explosives, although many firefighters and residents who confronted the blaze remain concerned about the potential exposure to various radioactive elements released by the fire.

The fire in town burned 239 homes. The prescribed burn that was intended to scorch about 900 acres over a year or more's time instead torched 48,000 in a matter of days. It cost $1 billion to fight and recover from, making it the most expensive U.S. forest fire to that date.

Miraculously, as opposed to the Lower North Fork Fire that grew out of a similar controlled fire in Colorado 12 years later, nobody perished in the Cerro Grande Fire. The lab reported that no radioactive material escaped during the blaze. But there was plenty of fallout.

ROGER KENNEDY HAD RETIRED AS director of the National Park Service three years before the Cerro Grande Fire, but he watched it from his home east of Los Alamos. The blame for the fire landed squarely on some of his former employees — the ones who'd started the prescribed burn. "I saw people that were doing very hard work hung out to dry," he told me a few months before cancer took his life in 2011.

In his storied career Kennedy had worked in five presidential administrations, was a White House correspondent for NBC, hosted his own Discovery Channel series, and ran the Smithsonian's National Museum of American History. He wrote more than a dozen books.

As investigators interviewed everyone involved in starting and fighting the fire, Kennedy dug through more than half a century of history. Nuclear threats, he discovered, had more influence on the United States' relationship with wildfire than any of the firefighters in Los Alamos could have realized.

"The story," he told me, "... extends to the entire arrogance of the scientific community after the development of the bomb. They didn't understand what they were doing outside of the bomb."

During World War II, newsreels and magazines showed European cities in rubble. Kennedy told me about a book from the National Fire Protection Association that showed photos of charred bodies and a map of "United States cities corresponding approximately to destroyed cities of Germany and Japan." After seeing the devastation of Hiroshima and Nagasaki, Edward Teller, father of the hydrogen bomb, worried that American cities were vulnerable to similar destruction.

"Teller feared atomic annihilation," Kennedy told me. "You couldn't grow up without images of Dresden, Rotterdam, Hiroshima, Nagasaki — the destruction of great cities by modern war."

Teller, who had some experience with urban planning from his involvement with "atomic towns" such as Los Alamos, promoted a government program to move the population of American cities to rural regions. "In an atomic war, congested cities would become death-traps," Teller wrote in a 1946 edition of the *Bulletin of the Atomic Scientists*. "Dispersal of cities may mean the difference between extermination of one third of our population and the death of only a few million people . . . We cannot do this without government regulations, just as restrictive or more restrictive than those of the recent war years."[4]

Other nuclear scientists followed suit. "Dispersal is the only measure which could make an atomic 'super Pearl Harbor' impossible," wrote Eugene Rabinowitch, editor of the *Bulletin*, in one of a number of articles that promoted dismantling the nation's "concentration of attractive targets."[5]

After these articles appeared, Harry Truman called for "general standards with respect to dispersal, . . . in the allocation of critical materials for construction purposes, and in the making of emergency loans."[6]

Plans for tax-subsidized dispersal included Federal Housing Administration loans and GI insurance, federal slum clearance loans, aid for highway construction, loans to industry, defense contracts, and tax concessions for those who moved from dense cities to what 50 years later would be known to firefighters as the wildland-urban interface. "People were induced by immense subsidies to find themselves homes as far away from industrial sites as possible — and thus some of them wound up in danger of another kind of fire," Kennedy wrote in his 2007 book, *Wildfire and Americans*.[7]

Racial fears pushed population from some cities, while the mythology of the abundant West pulled people to the woods. The weather eased the path for the nation's exodus. The 1950s and '60s were unusually wet and cool decades in normally hot and arid western states. That made the grass greener, and less black, on the west side of the fence dividing urban and wild lands.

Still, federal policy drove the migration. "It is wrong to believe that postwar American suburbanization prevailed because the public chose it and will continue to prevail until the public changes its preferences," wrote professor of social work and urban planning Barry Checkoway. "Suburbanization prevailed because of the decisions of large operators and powerful economic institutions supported by federal government programmes, and ordinary consumers had little real choice in the basic pattern that resulted."[8]

President Dwight Eisenhower's Federal-Aid Highway Act of 1954 allotted $175 million for the interstate system. That grew into the Highway Trust Fund, which spent more than $700 billion building highways over the next 30 years. The migration supported myriad industries: the building of homes and roads, power grids, dams, and water diversions. Shopping malls were both hearts of commerce and evacuation centers. The patriotic strategy turned into an economic engine. "Dispersal is good business," Tracy Augur, of the American Institute of Planners, boasted to readers of the *Bulletin of the Atomic Scientists*.[9]

Congressional representatives in rural states competed for the industry and development leaving the cities. The migration was soon driven more by barrels of pork than bombs of plutonium. "Billions of federal dollars were spent to induce millions of Americans to migrate out of cities and along highways into the suburbs and beyond — to the edges of natural systems that did not appear at the time to be as dangerous as they turned out to be," Kennedy wrote.

During the 50 years prior to the Cerro Grande Fire, Kennedy told me, one-fifth of the American population moved into the nation's most flammable forests, chaparral, and grasslands. He described Boulder, where I'd just moved, as the "most fire-endangered city on the Colorado Front Range." Boulder County's population had grown from 48,000 to 278,000 in the second half of the twentieth century.

"Thousands of people are moving unwarned into firetraps every day," he wrote. "Indeed, they are being encouraged to settle there by taxpayer subsidies."

Those perverse financial incentives continue today. County governments benefit from the tax base of new developments, but when a big wildfire threatens those communities, it's the federal taxpayer at large who pays for their protection. So city dwellers who face no threat of wildfire subsidize people to live in the country's most flammable landscapes.

I asked Kennedy what would stop people from moving into the path of wildfires. "A lot of fires, a lot of death, and a lot of painful learning," he responded.

• • •

WHILE HUMAN SETTLEMENTS CAN ENCROACH ON or retreat from Red Zones in a matter of years, nuclear leftovers linger for centuries.

In Ukraine, the Red Forest around Chernobyl is named for the rusty color the pines took on after radiation killed them. Plutonium, strontium, cesium, and other fallout from the 1986 explosion and fire in the nuclear power plant contaminate the woods. In 2011 researchers from Europe and the United States determined that a wildfire in the forests of Chernobyl would send that fallout back into the atmosphere, where it could spread far beyond the exclusion zone.[10]

And the risk of a large wildfire there grows yearly, reported Sergiy Zibtsev, the lead author of the 2011 paper and a Ukrainian forestry professor who has been studying the Red Forest for two decades. "In Ukraine, people just don't realize how dangerous [big wildfires can be]," he said in an interview. "They have [the] completely wrong approach in the exclusion zone. They [think] that they can manage any fire, and this is a very, very big mistake."

After the population moved from the "zone of alienation" and commercial activities were forbidden within it, the woods grew thick and tangled. Birds in the exclusion zone have smaller brains, trees show stunted growth, and wild boars as far away as Germany have dangerously elevated radiation levels.[11] Radioactivity killed off many of the bacteria, fungi, and invertebrates, slowing the decomposition of needles, leaves, and trees by up to 40 percent.[12] Those fuels have accumulated.

In the meantime, persistent droughts, longer summers, and decreased rainfall have primed the Red Forest to burn. "Here everybody [is] sure that we [are] already living in a different climate," Zibtsev said.

After witnessing massive U.S. wildfires in 2005, Zibtsev grew concerned that a blaze in the Red Forest could create a nuclear disaster. Thinning the forest and preparing for fire could reduce the risk, but convincing the Ukrainian government of the threat proved difficult.

"It came back to that same old issue," Chad Oliver, director of Yale's Global Institute of Sustainable Forestry, said after visiting Zibtsev in the Red Forest and collaborating with him on the 2011 paper. "What if you knew something horrible was going to happen and couldn't get anyone to listen to you?"

Oliver told his host, "These could burn any minute."

Research by Zibtsev and Oliver showed that a fire that fully consumed the forest would blanket Kiev with radioactive smoke, leading to an increased risk of cancer there. Agricultural products up to 90 miles away from the fire would be so contaminated they couldn't be safely consumed. And the stigma of radiation on the nation's farms would keep other countries from importing even uncontaminated Ukrainian foods.

"The estimated cancer incidents and fatalities are expected to be comparable to those predicted for Fukushima," a team of scientists wrote in 2014, noting that a wildfire in the Red Forest would rate high on the International Nuclear and Radiological Event Scale.[13]

A Red Forest megafire would expose firefighters to radiation both externally when the flames released it and internally through the smoke they inhaled. That's on top of the peril they would face battling a blaze they are neither equipped nor trained to fight. The entire exclusion zone has only six fire towers and just one helicopter that doesn't even drop water.

An early warning system, modern firefighting equipment, and forest thinning would reduce the hazard, Oliver said. But those options are expensive.

In the summer of 2014, Ukraine expanded the exclusion zone, and the United Nations earmarked funds to develop a wildlife preserve there. But with the nation engaged in a civil war, the possibility that Ukraine would do much to reduce the nuclear threat of the Red Forest seemed remote.

ON APRIL 28, 2015, two days after the 29-year anniversary of the Chernobyl disaster, an out-of-control wildfire spread into the exclusion zone, burning down an abandoned village and extending to within 10 miles of the mothballed nuclear plant. The fire didn't burn into the most contaminated forest, and Ukrainian scientists measured only a slight increase in radiation. Two months later wildfires burned into the heart of the Red Forest. This time Ukrainian nuclear inspectors detected radiation of approximately 10 times normal levels.

"It's like Chernobyl all over again," Chris Busby of the European

Committee on Radiation Risks said, describing the potential that the fires would send radionuclides absorbed by the trees back into the air.[14]

And from Japan to the Rocky Mountains, other wildlands were growing red.

IN MY HOMETOWN THERE WERE fears of fallout from less remote smoke.

Before it was raided by the FBI in 1989 and closed due to a litany of covered-up safety violations and crimes, the Rocky Flats Plant, eight miles south of Boulder, made small plutonium bombs that were the triggers for hydrogen bombs. During the plant's operation, fires, accidents, and routine work dusted the surrounding land with radioactive particles. In the 1970s a federal government study found plutonium particles around the 10-acre site of the plant, and the Atomic Energy Commission found them throughout the Denver metro area. Cleaning up the Superfund site took 10 years and cost nearly $7 billion. The facility's buildings were razed, and the U.S. Fish and Wildlife Service took over 80 percent of the sprawling grassland surrounding them and turned it into a wildlife refuge.

Eons of erosion from the nearby mountains covered the rocky mesa with a thin layer of soil that supports Colorado's largest patch of dry tallgrass prairie, which contains smaller versions of the same species as the moister prairies of the plains states. Like that grassland, Colorado's tallgrass is dependent on fire. But after the creation of the Rocky Flats Plant, no natural fires were allowed on Rocky Flats. Trees and invasive species encroached. The soil became anemic from the lack of flame-released nutrients. Animal species dependent on the fire-nourished grasses fell into decline.

Dave Lucas, the manager of the refuge, carries a Geiger counter whenever he travels through the grassland. He told me he has never measured dangerous levels of radiation, but sees impacts from the lack of fire every day. "Grassland birds are the fastest-declining species," he said.

Mammals, like the threatened Preble's meadow jumping mouse, are also slipping away. As we walked in the grass, Lucas pointed out a fabulously purple penstemon flower that would be more common, along with many others, if the grassland burned.

"It's a fire-adapted, fire-dependent landscape," he said. "You take any major factor out of play and you get a different ecosystem. Fire is a natural part of this habitat."

In 2014 the Fish and Wildlife Service, in an effort to save the tallgrass, planned a prescribed burn on 701 acres of Rocky Flats. A dozen years earlier a 50-acre test burn spread smoke from the mountains to Denver. Although federal and state agencies noted no dangerous levels of radiation within it, a local antinuclear activist, Paula Elofson-Gardine, claims to have measured radiation up to 1,300 times the normal background levels.[15]

Homes were sprouting next to Rocky Flats in 2014. Candelas, a nearly 1,500-acre planned community, was still under construction when the burn was announced. All of its 2,500 homes would be within a mile of the site of the former nuclear facility, and many abutted the refuge. Although tests showed the grassland held minimal radiation and the fire's smoke would be below federal limits for radionuclides, opposition to the prescribed burn exploded among both the new residents and antinuclear activists once aligned against the bomb plant.

The Fish and Wildlife Service canceled the burn.

Lucas welcomed Rocky Flats' next-door neighbors as a population that actually cares about the refuge, but noted that some of their homes are within the reach of wind-driven flames from the prairie fires that become more likely each year the grassland doesn't burn. Grazing or mowing the grass could reduce the hazard but won't save the tallgrass, or the species dependent on it.

The USFWS still hopes to hold controlled fires at Rocky Flats. One way or another, Lucas said, the grassland is certain to burn.

15

THE VANISHING FOREST

WHILE RADIOACTIVE FALLOUT AROUND CHERNOBYL led to a thickening of the forest, the woodlands around Los Alamos were getting "nuked" by wildfires.

Valles Caldera was named a national preserve in 2000, the year the Cerro Grande Fire burned hundreds of homes in the nuclear city just east of it. Eleven years later winds blew an aspen onto a power line at the edge of the caldera. In its first 14 hours the Las Conchas Fire burned 43,000 acres — as much as the Cerro Grande burned in two weeks. In total the 2011 fire destroyed 63 homes and 49 outbuildings, and grew to 156,593 acres — almost three times the size of its predecessor.[1]

At its peak it burned nearly an acre a second.[2] Its fastest rates of spread were downhill and against the wind — behaviors never seen before by the firefighters in front of it. It created its own weather and sent another massive plume of smoke into the sky over Los Alamos.[3] This time officials wasted little time evacuating both the lab and the city.

"I think it's been 8,000 years since we've seen a fire of this severity," Craig Allen told me when I visited Valles Caldera after the fire.

"The really ugly stuff goes on for 10 miles that way with barely a tree standing," he said, pointing to the southeast. "There's a 30,000-acre hole in what used to be forest and woodland, centered on that first day's run."

Allen, a research ecologist with the U.S. Geological Survey, was in his 30th year working in the Jemez Mountains when I visited. He pointed out a six-square-mile area where the blaze killed almost every tree.

The term "moonscape," like "megafire," is bandied about by some in the fire community and disdained by others, but in the case of the Las Conchas Fire, it was hard to think of a more accurate description. At the heart of the wasteland, the forest was "nuked"—another firefighter's term that's eschewed by science but fitting for both the scene and the atomic laboratory down the road. A charred stubble of stumps spread for miles on ground burnt bare and brown. The lone black skeleton of a 30-foot alligator juniper stood in the middle. Thousands of incinerated ponderosas looked like brushstrokes of black ink against the gray ash. Other pines were dead but not charred. The fire was so hot that it killed them without touching them.

Heat had killed more than 90 percent of the preserve's piñon pines before the fire even started. Tree rings show that the dry spell after the turn of the millennium was less severe than others the piñons survived in the past, but temperatures ran a couple of degrees hotter. Studies by Dave Breshears, an arid lands ecologist at the University of Arizona, and Nate McDowell, a tree physiologist at the Los Alamos Lab, showed that the increased heat magnified the impact of the drought, making the trees more vulnerable to insect infestations. The bugs ultimately killed them, but the piñons could have spit them out if the temperatures hadn't weakened them or the drought hadn't deprived them of the moisture they need to defend themselves with sap.[4]

McDowell tested how much heat and dryness the trees can endure by enclosing piñons and junipers in acrylic cylinders (which look like giant test tubes from a science fiction movie) in which he can control the temperature and moisture. The trees around Los Alamos, his experiments showed, are at a climatic tipping point. They can tolerate little more drought and heat, and those aren't the only limits they are pushing up against.

Craig Allen and Bob Parmenter, the science and education director for the Valles Caldera National Preserve, led me to a slope that looked down over the extinct volcano. Grassy meadows filled the bottom of the crater, while ponderosa pines, junipers, and a few surviving piñons covered its slopes. "Grasses dominate the caldera," Allen said. "With a reverse tree line."

The forest is plagued both by the death of some trees—the piñons—and an overabundance of others.

Historically, the woods took to the colder, moister soils above the meadows. Grasses and the ground fires they fueled kept the trees out until grazing livestock arrived. More than 100,000 sheep dined on the caldera's grasses early in the twentieth century, followed by as many as 12,000 cattle after World War II. As the livestock grazed down the grasses, the pines moved in. Using aerial photos, Parmenter determined that the area of open grassland declined by 55 percent between 1935 and 1979.

Decades of grazing and fire suppression left some of the forests in the caldera with an exponentially greater density of ponderosa pines and understory growth. The excessive fuel turned mellow ground fires into explosive crown fires, and the thirsty vegetation magnified the water shortage in the already arid woodlands.

The preserve has since reduced the livestock in the caldera by 95 percent, but it takes a long time to fix a forest. And there's no way to mitigate the increasing temperatures and drought plaguing Los Alamos.

Thirst and bugs can wipe out a forest, but they don't kill every tree. Young trees that survive will eventually recolonize the landscape.

But megafires are different. "When you have a high-severity crown fire burn through a piñon stand or ponderosa stand over a huge area, it takes out all the seed source," Allen told me. "That's why these . . . big, high-severity fire patches are a big deal. We're not getting forests back."

While even Allen would admit that good-sized patches of severely burned forests are an important part of many forest fire mosaics, in some fires those areas are so large that they remain treeless. Trees that release their seeds in wildfires to resurrect a forest — called obligate seeders — are vulnerable to an increased frequency of fires. Blazes that recur more often than a forest has adapted to can kill the young trees regenerating from the previous blaze before they're old enough to drop their own seeds, thus wiping out the species in that location unless the area is small enough that unburnt trees nearby can spread seeds to the burnt area. In some cases a new forest of trees better suited to the increased fire frequency will grow, but in many regions shrubs, weeds, or grasses replace the trees. Research published in 2013 by David Bowman at the University of Tasmania, in Australia, shows that abrupt changes in the frequency or intensity of fires, like those around Los Alamos, can bring landscape-wide loss of obligate seeder forests.[5]

"If you don't have any mother trees surviving in the midst of these big patches, you don't get tree regeneration," Allen said. "You have a lot of time for shrubs and grasses to take over . . . which then makes it more difficult for trees' seedlings to establish, even if there were seeds."

In multiple lines of evidence over thousands of years, Allen sees human impacts having something of a slingshot effect on the fires and forests in the Jemez Mountains. For at least 6,000 years frequent low-intensity ground fires shaped the landscape — keeping forests thinned and preventing timber from overtaking grasslands. But with the arrival of railroads, grazing livestock, and, finally, the nuclear laboratory and the town that served it, humans snuffed that fire cycle. The forests grew thick and sick.

The Cerro Grande Fire that burned into Los Alamos marked the snapping of the rubber band. During the subsequent 15 years a cycle of intense drought and record heat fell onto the overgrown forests. Crown fires in New Mexico and Arizona burned bigger now than in hundreds of years[6] and proved nearly impossible to stop. "At the Cerro Grande, the fire model shows that if we hadn't fought it at all, it would have gone exactly the same way," Allen said.

During a severe drought that struck the area in the 1950s, the largest fire in the southwestern United States was about 50,000 acres, according to Tom Swetnam at the University of Arizona's Laboratory of Tree-Ring Research. Today the Southwest is seeing fires nearly 10 times that size. In Arizona the Rodeo-Chediski Fire burned 468,000 acres in 2002, and the Wallow Fire burned 469,000 acres just nine years later. It's a phenomenon so well documented in the region that it has come to be called the Southwest Ponderosa Pine Model.

Swetnam and fire anthropologist Christopher I. Roos looked at nearly 1,500 years of fire scars and tree rings to create a statistical analysis showing that during both the warm centuries and the cool ones, ground fires behaved largely the same until this century. Recent megafires in the Southwest, their research suggests, would not be occurring if humans hadn't removed ground fires, added grazing animals, and otherwise impacted the landscape.[7]

The forests survived deeper droughts in the past. But the combination of drought, increasing temperatures, overgrown woods, and overgrazed grasslands drives fires that overwhelm their resilience. The size of patches where nearly every tree is killed — like that 30,000 acres in the Las Conchas Fire — leaves mother trees too far away to provide the seeds that would regenerate the forest.

The impacts trickle down to plant, animal, and human communities. After the Las Conchas Fire, the cities of Albuquerque and Santa Fe had to close their water intakes on the Rio Grande for some 40 days due to contamination and gunk from the fire. That has led those cities and organizations like the Nature Conservancy to radically scale up the amount of prescribed burning and thinning in the woodland watersheds.[8]

The threats to the forests of the Southwest are increasing rapidly. In the fall of 2012 researcher A. Park Williams and colleagues created the "forest drought-stress index" using more than 1,000 years of tree-ring data from ponderosa pines, piñons, and Douglas firs — the dominant conifers in the Southwest.[9]

The current drought is "the worst . . . in the last thousand years," Craig Allen, one of Williams's collaborators, told me. And Williams's index shows that if climate models are correct, the droughts of the coming century will continue to exceed any in the past millennium, dealing even more punishment to trees. In his 2013 paper he describes how "vapor-pressure deficit" stresses them. The atmosphere around the trees is so dry and hot that the stomas — the tubes that move water through trees — close to keep the plant from drying out completely. But the trees, which also get their nutrients through the stomas, will starve if the tubes remain closed for long. And if the hot, dry air draws moisture from the trees faster than their circulatory systems can move it, the stomas can collapse under the stress like straws sucking ice cream.

By 2050, Williams contends, conditions will be beyond what most coniferous forests in the region can survive.

"Even without climate change, [which is] expected to drive further stress to forests and to change tree species and to change where they are on the landscape . . . we're talking centuries without trees," Allen said. "That 30,000-acre hole, most of that will still be treeless several centuries from now . . . tens of thousands of acres of Las Conchas that won't be back as forests for centuries."

The decline of those forests, Allen noted, may be a harbinger of widespread forest changes around the planet.

Through heat, thirst, and the twin plagues of insects and wildfire, the warming climate is driving forest die-offs on every wooded continent. In California, Forest Service researchers in 2016 documented 102 million trees killed by drought and insects across the state since 2010, 62 million of which died just that year.[10] The Carnegie Institution for Science identified tens of millions more that were suffering severe enough drought stress that they were at risk of dying.[11]

Already many forests that were once carbon sinks are releasing more CO_2 than they suck from the atmosphere as they fill with unhealthy, old, or unnaturally fire-prone trees. Woods that once cooled the earth are beginning to warm it, putting the forests around Los Alamos in the middle of a vicious cycle of climate and fire.[12]

LATE IN THE SPRING OF 2012 I climbed a peak above Boulder hoping to find blue sky, but the air was filled with enough smoke to make me cough. The Whitewater-Baldy Fire burning in the Gila Wilderness of New Mexico, more than 600 miles away, had grown into the largest fire in the state's history — surpassing the record set by the Las Conchas Fire just the year before. It was hard to see hope in the haze, but it was there when I looked for it.

Jose "Pepe" Iniguez, a research ecologist with the U.S. Forest Service's Rocky Mountain Research Station, was eager to get into the Gila Wilderness (part of the Gila National Forest) after it burned. He had plots set up to compare how natural fires, prescribed burns, and fires extinguished by firefighters affected the behavior of future fires. Many of his study plots were in areas overrun by the Whitewater-Baldy Fire. But when he revisited them after the blaze, he didn't see a nuked forest. "To me, it's the best-looking forest in the Southwest," he said. "There's a lot [fewer] trees, and the trees that are growing are big."

In his plots of ponderosa pines, the Whitewater-Baldy Fire — actually two fires that ignited from lightning strikes and then merged — burned away brush and grasses on the ground and killed smaller trees that crowded the woods. On the big pines that remained, the flames burned off the low branches that could carry a future fire from the ground into the treetops. As he walked through the stands opened up by the fire, Iniguez could see scorch marks 8 to 10 feet up the thick trunks of the surviving trees, which stretched into the sky like widely spaced columns. Black char on the base of the trunks made red bark higher up seem to glow. The forest canopy they held up arched some 40 feet over the researcher's head.

"When you're in there, you can look and you can see a long ways," he said. "That's one of the cool things about the Gilas. You're in a pine forest, but all you see is stems."

The parklike condition of his study plots isn't just the result of the 2012 fire but points to the success of almost 40 years of letting fires run their natural course in the Gila—the nation's first designated wilderness area, and the first to let wildfires burn when they didn't threaten private property or vital resources. Where previous fires had cleared out smaller trees and brush, the Whitewater-Baldy burned with less intensity, leaving the larger, more resilient trees. In pockets where previous fires had been extinguished, the accumulated timber, brush, and grasses drove explosive fire behaviors that killed virtually every tree. At higher elevations, crown fires during the Whitewater-Baldy leveled ancient spruce-fir forests high on the Mogollon Rim, which were among the Gila's most dramatic vistas. Those forests burned in patches twice as large as they had in recent history. But at lower elevations, the results of previous fires calmed the massive blaze.

"All of our plots in the ponderosa pine forests did just fine," Iniguez said. "Some of those areas have burned four times."

Ecologists call this type of fire "self-regulating." According to Iniguez, "As soon as it hit the ponderosa pine belt, it went from a crown fire to a ground fire."

The ground fires burned with far lower intensity than the crown fires. In the ponderosa forests they left a more natural density of a few dozen widely spaced trees per acre, as opposed to as many as 1,000 crowding acres where fires had been extinguished for decades.

Only 13 percent of the Whitewater-Baldy Fire burned hot enough to kill all the trees in an area. That's about half the proportion that had burned with tree-killing intensity in the Las Conchas Fire the year before.[13] The lower proportion of trees killed relates to another tiny percentage change. In the Gila, firefighters fight about 98 percent of lightning-ignited wildfires, while in the nation overall that figure is 99 percent. That 1 percent reduction allows about 100,000 acres a year to burn in natural wildfires, on top of some 20,000 acres of prescribed burns set by forest managers.[14]

In 1995 U.S. land agencies such as the Forest Service, the Park Service, and the Bureau of Land Management, recognizing that low-intensity fires are good for the forests, agreed to reintroduce fire to remote wildernesses like the Gila. The cheapest and most efficient way to achieve that goal is to allow small lightning fires that seem unlikely to grow big or threaten property to burn.

But convincing the American public — which had spent a century learning to hunt down every fire in the woods — to let these fires run wild would be more difficult than fighting the blazes.

16

THE FIRE-INDUSTRIAL COMPLEX

Washington, D.C. — May 25, 2012

A PIECE OF PAPER CAN IGNITE as big a fire as a lightning strike, particularly when the paper is from Washington.

Despite the improved health of many forests where natural fires burned, two weeks after the fire in the Gila Wilderness ignited, a memo from James E. Hubbard, a deputy chief in the U.S. Forest Service, ordered "aggressive initial attack," including on fires in the most remote wildernesses.[1] Even small ground fires that improved wildlife habitat, cleaned up watersheds, and reduced the risk of big crown fires would be snuffed. The order outraged ecologists and firefighters both within and outside the Forest Service who had worked for decades to bring natural wildfires back to America's wildernesses.

Hubbard's reasoning highlighted an underlying irony of wildfire management. Putting all fires out when they are small is cheaper than risking the cost of one of them growing into a big "project fire."

The Forest Service had exhausted its funding to fight wildfires during most years of the previous decade, and 2012 was forecast to break the budget again. Funding is based on the average of the past 10 years of firefighting costs. But with annual costs rising steeply as fire seasons became more destructive, the average trailed the yearly price tag for firefighting, and the Forest Service was virtually guaranteed to run out of funds to fight wildfires, as it had done in half the fire seasons since 2000. Putting out every fire, even remote or beneficial ones, would save money.

"Safe aggressive initial attack is often the best suppression strategy to keep unwanted wildfires small and costs down," Hubbard wrote in his memo.

Wildfires went from taking up 16 percent of the U.S. Forest Service budget in 1995 to more than half of it 20 years later.[2] They also consumed more than 10 percent of all Department of the Interior agency budgets, cost state governments up to $2 billion a year, and resulted in almost incalculable costs to local and county governments.

But while the cost of wildfires drew a trend line that climbed as steeply as the Rocky Mountains, politics and economics brought cuts to government programs to fight, prevent, prepare for, and recover from forest fires. Between 2010 and 2012, the federal budget to deal with wildfires had dropped by $512 million, or about 15 percent. In August 2013 the Forest Service ran out of money to fight forest fires just as the nation reached its highest level of wildfire threat, forcing it to divert $600 million from other programs, including those to reduce hazardous fuels, in order to continue chasing smoke. The feds had already cut funding to thin dangerously overgrown forests by nearly a quarter of the previous year's budget.[3]

"A person has to wonder. Is this going to be the new norm—frequent record-setting fires, while the number of federal firefighters and air tankers continue to shrink?" Bill Gabbert, a former fire management officer and hotshot, wrote on his website, Wildfire Today.[4]

A 2009 federal review of the nation's wildfire policy noted that government agencies overshot their firefighting budgets in each of the five previous years due to the increasing size and destructiveness of wildfires. Between 2002 and 2014, the Forest Service and the Department of the Interior diverted $3.2 billion from research, forest health, and recreation to fighting fires.[5]

Counterintuitively, the government's solution to the decreasing funds available to battle wildfires is to fight more fires. "I acknowledge this is not a desirable approach in the long run," Hubbard admitted in his memo.

For many, his order to fight fires in wilderness areas wasn't a desirable approach in the short run either. Resources sent to blazes in remote forests would be difficult to redeploy to fires that threatened communities and infrastructure. And some of the blazes they extinguished were likely to beget more serious fires later, while drawing desperately needed funding and resources away from other critical programs that prepare for future fires.

"We have a choice on how we receive the inevitable smoke from Montana's fires," wrote Bob Mutch, a retired Forest Service fire manager with 40 years of experience, in the *Missoulian,* "in smaller, regular doses over time, the result of free-burning fires in wilderness and sound forest management practices outside wilderness, or in supersized doses from megafires that are the result of fire exclusion, unnatural fuel accumulations, and a changing climate."[6]

FOR OTHERS, HUBBARD'S ORDER didn't go far enough. Some pointed incriminating fingers at the Whitewater-Baldy Fire, which, although it improved the health of the parts of the Gila Wilderness where Pepe Iniguez had his study plots, burned nearly 300,000 acres, destroyed a dozen cabins, and forced the evacuation of several towns.

"As I toured the Whitewater-Baldy fire, a Forest Service Fire Manager told me, 'This forest will not grow back for 100 years.' How is this 'managing our forests'?" Steve Pearce, the Republican representative for New Mexico's second district, wrote to me.

"The policy of 'let it burn' only works when nature is in balance," he continued. "But currently, it is not. After the fire, tons of ash [washed] into the streams, killing endangered species. How is this approach ecologically sound?"

As the Whitewater-Baldy Fire burned, Pearce took to the floor of the U.S. House to make a blistering attack on the policies that had allowed the fire to run its course in his district. "It's a tragedy, what's going on in the most pristine parts of our country," Pearce said, "wilderness areas where fields have been allowed to burn and where we're going to see . . . absolute destruction."[7]

As opposed to the forest cathedrals that Iniguez saw after the blaze, Pearce cited soils cooked almost into glass and debris flows that would "flood towns completely off the face of the Earth."

Like Iniguez, Pearce believes that western woodlands are overgrown. But instead of letting wildfires clear out the woods, he thinks that the forests need thinning by logging companies and that flammable grasses should be mowed down by grazing animals. He urges a return to the Forest Service's firefighting policy of the 1930s.

"In visiting with the head of the U.S. Forest Service this week, I asked about a policy that used to exist to put out fires . . . the 10 a.m. policy," Pearce said. "That is, if we see a fire running at any time today, we're going to put it out by 10 a.m. tomorrow; and if we don't get it out by 10 a.m. tomorrow, we're going to put it out by 10 a.m. the next day. I want you to go back to the 10 a.m. policy."

Foresters and firefighters, however, have long cited the 10 a.m. policy — which grew out of the response to the Big Blowup of 1910 and for nearly 40 years saw every natural wildfire ordered extinguished by the morning after it was first seen — as a major cause of many of our nation's forests growing explosively dense in the first place. "I think the whole idea of the 10 o'clock policy isn't going to work," Iniguez told me.

When the policy was in place, from the 1930s into the 1970s, firefighters took risks that aren't tolerated today, he said. And for much of that time, western forests, aside from regional droughts like the one in the Jemez Mountains in the 1950s, were in a cycle of cooler and wetter weather that helped extinguish fires.

In their polarized views of how to restore health to the nation's forests — letting fires burn as opposed to putting them out fast — Iniguez and Pearce represent two camps battling in the ash of the past decade's epic fire seasons.

On one side, fire ecologists see flames as being as much a part of the forests as the trees, and believe that allowing natural fires to burn will heal those forests and keep them healthy. Firefighting, they say, should focus on protecting homes, watersheds, and infrastructure, while blazes in more remote woodlands should be allowed to run their course.

On the other side of the debate, an alliance often called the "fire-industrial complex" believes that timber harvests and grazing animals can thin the woods, while investments in more aggressive firefighting and bigger and better technologies can protect the valuable resources that related industries depend on, as well as the communities spreading fast into the forests.

Few firefighters or foresters deny the impact of the current warming and drying climate in driving the increase in western wildfires. Some also note that the wetter and cooler climate during much of the previous century fed the crisis by breeding overconfidence in firefighters' ability to extinguish wildfires. This has led policy makers to believe that with more firefighters, improved technologies, and more equipment, we can once again extinguish every forest fire the day after it ignites.

"The time when we were successful at doing fire suppression, from the 1930s to the 1980s, it's not really comparable to the current situation," George Wuerthner, an ecologist who edited the book *Wildfire: A Century of Failed Forest Policy,* told me. "Even with more modern equipment, with airplanes and helicopters, we can't stop the fires."

Indeed, while the nation's investment in fighting wildfires exploded, so did the fires. "I've had firefighters tell me, 'It's like dumping dollars on the fire,'" Wuerthner said.

While judicious logging and grazing can reduce the amount of fuel available to burn in the forests, they can also increase fire activity. Timber interests prefer to harvest large trees, which are more profitable, but these are the very trees that make forests resilient to wildfire. Small trees and brush, the removal of which improves the health and lowers the flammability of the forests, hold little value. Debris from logging activities, known as slash, can fuel fires if it isn't cleaned up properly. New logging roads allow more people into the woods, which leads to more campfires, more sparks from vehicles, and more flashes from firearms to ignite blazes. "The more logging roads you have, the more starts you have," Wuerthner said.

Even ecologically sound logging and thinning can increase the potential for fires. "When you reduce the density of the trees, the shrubs and grasses fill the gaps," Wuerthner said. "Logging will increase evaporation. Logging roads will allow wind to penetrate the forest."

If increased logging, grazing, and firefighting efforts often result in more destructive fires, why does every big wildfire season bring new calls to send more air tankers, loggers, and cattle onto public lands? For Wuerthner and many other fire ecologists, the answer is simple: money.

Rich Fairbanks, a former fire planner for the U.S. Forest Service and onetime foreman of a hotshot crew, points out that while most forest firefighting in the past was done by federal employees, today the private sector does much of it. "Privatization has changed firefighting, and not for the better," he said.

Private companies supply everything from firefighters and bulldozers to caterers and mobile shower facilities for the fire camps. Most don't get paid if they're not actively fighting a fire, so they lobby to fight as many fires as they can. Today the private sector provides about 40 percent of the nation's wildfire-fighting resources.[8]

"There are a lot of people who see a return to a 10 a.m. policy as a way to make money," Fairbanks said. "Almost all of the helicopter forces are contractors. Some of them do have money, and some of them do get involved in politics."

Developers building what Fairbanks calls "suicide subdivisions" also push for increased firefighting so they can build deeper into dangerously fire-prone landscapes and continue using flammable building materials. "That's one of the uncounted costs that developers get away with," Wuerthner said. "It's because county commissioners won't say 'no' to any developer."

Helicopters working wildfires cost up to $35,000 a day on standby, plus around $7,000 an hour for flight time, while air tankers run around $10,000 a day and as much as $5,000 an hour when in the air. But much of the expense of a wildfire is just for getting the fire camp up and running. Cattlemen's Meat Company (with the slogan "We Feed on Disasters") and For Stars Catering (which serves Hollywood film productions as well as firefighting camps) can feed 2,500 firefighters at a cost of more than $100,000 a day when they are running at capacity. A 12-stall shower trailer for fire crews to scrub down in runs $4,735 a week. Mobile laundries charge as much as $500 an hour.

"There will be an increasing polarization of this debate," Crystal Kolden, an assistant professor of geography at the University of Idaho who focuses on wildfire, told me. "Science suggests that we should let more of these fires burn, particularly in wilderness areas. Fires become self-regulating over time. Older fires are a firebreak to newer fires."

But the terror that large wildfires inspire, she said, allows profit to trump science. "There's a lot of money to be made in fire suppression. Those private entities . . . have an incredibly powerful lobby in

Washington, D.C. Contractors are very good at playing on the public's fear of large wildfires."

Nowhere is the debate more heated than among current and former employees of the U.S. Forest Service.

"I'm actually appalled that, after all of our experience over about 100 years, [Steve Pearce] would have even considered a 10 a.m. policy," said George Weldon of Missoula, Montana, who was deputy director of fire and aviation for the Forest Service's northern region when he retired in 2010.

"I'm just amazed in the ... years since I quit, how they've moved backwards," he said of the Forest Service's fire policies. He told me that he saw marked improvement in the health of the Selway-Bitterroot and Bob Marshall Wildernesses of the northern Rocky Mountains, where he worked, and that he was outraged when the Forest Service ordered suppression of fires deep in those remote forests.

Weldon cited air tankers, which he helped manage when he was fighting fires, as an example of how industry drives wildfire policy. "The Forest Service is looking at spending $500 million to get new-generation air tankers," he said. "But there's never been a scientific study that demonstrates the effectiveness of large air tankers. If a study was ever done, in my opinion, they are not worth the money at all."

Weldon and other retired Forest Service firefighters cited estimates that air tankers are effective only about 30 percent of the time. "It's a prime example of how powerful the fire-industrial complex has become in a very short time," he said.

Weldon believes that the Forest Service and other federal agencies involved with firefighting present wildfires as disasters because they need to make up for budget cuts and loss of revenue from the decline of timber sales. "It's not only the politicians and industry. It's the agency," he said. "They're trying to keep their agencies alive, and the way to do it is to play on people's fears."

Yet, of the half-dozen reports that Congressman Pearce has on his website to support a return to the 10 a.m. policy, the one prepared for him by another retired Forest Service leader is the most pointed in its attack on letting fires burn.

Bill Derr, a retired special agent in charge of law enforcement on

California's Forest Service land, wrote that the "ambiguous" national fire policy contributes to the increase in wildfires and their damage. "The elimination of the time-honored 10 AM Control Policy," he wrote, prevents the Forest Service from containing fires early and makes them more costly. "Allowing unplanned fires to burn has often resulted in escapes and more significantly created a belief by many Forest Service managers that fire is a positive change agent on the landscape thereby reducing their sense of urgency."[9]

Many fire ecologists, including Pepe Iniguez, told me how natural wildfires bring "positive change" to America's wildernesses, and I saw it for myself as I hiked through the Gila Wilderness, partially following the route that the Granite Mountain Hotshots had taken when they had come to fight the Whitewater-Baldy Fire with burnouts and chainsaws the year before they were burned over on Yarnell Hill.

I hiked through the vast burn zone of Arizona's Wallow Fire, the largest fire in Arizona history, which burned a month before the Las Conchas Fire, and then made my way into the Gila. For more than 100 miles, every forest I passed contained signs of recent fires. Ponderosa pine stands thinned with prescribed burns, natural fires, or timber operations showed blackened bases but green tops. Other, tighter stands looked like pincushions of black, 60-foot needles. An elk cow led her calf through the scorched trees. In the distance, the fire had left vast sections of the Mogollon Rim burnt of any color but black.

At the Glenwood Ranger District office I chatted with Shane Manning, a Forest Service engine captain who fought his first fire in the Gila when he was 12. "I threw a little dirt on it," he said.

Shane had worked for the Forest Service for 13 years when I stopped in. In just that time, he said, he'd seen a lot of changes. "Fire's year-round here now," he told me. "Most of our monsoons have been so dry."

The previous January he had been fighting fires in the wilderness. "You really have to be on your A game anymore," he said.

The first crew to fight the Whitewater-Baldy Fire thought they'd have it under control in three days. Six weeks and hundreds more firefighters later, they were still at it.

It took me a few hours to reach the heart of the burn in the Gila, where I walked through the nuked forest. First, as I turned in a circle,

I could see nothing but blackened, standing trunks. Farther in I wandered through an area where all that was left was cinders, ash, and charred trunks on the ground.

On my way out of the burn zone I stopped at the Gila Hotshots base and chatted with their superintendent, Dewey Rebbe. Dewey, tall with a gray ponytail, an earring, and the kind of lanky fitness that comes with years of chasing fires in the woods, lamented that while poorly managed logging and grazing could make forest fires worse, eliminating them altogether could do the same thing. "There's got to be a middle way somehow," he said.

Dewey brought up spotted owls, the poster child for environmentalists wishing to protect old-growth forests and, consequently, a scapegoat for the nation's struggling timber industry. In many areas where logging was banned to protect the owls, high-severity fires ended up driving them from their habitat anyway.

"You look at every owl pack when fire comes through, and they're gone," he said. In one forest he noted that four or five owl packs had abandoned the area after fires, he said. Of course, other birds, like blackbacked woodpeckers, sometimes took their place.

"Cutting trees is good for the forest," he said. "Cattle are good for the forest, in moderation."

Later, along a road in the forest, I noted a few dozen cattle grazing and, at one point, saw a patch of cinnamon fur beside one of them lope into the trees. When I climbed from my car, the cattle scattered, too. The bear and the cattle that hardly noticed each other but ran from me seemed to point to Dewey's "middle way," in which wilderness can exist beside our domesticated forests.

17

HIGH PARK

Fort Collins, Colorado — June 9, 2012

NOLAN DOESKEN WAS WORRIED, but he rarely let it show. After the Heartstrong Fire overran the Struckmeyer family in the waning days of winter, and the Colorado State Forest Service's prescribed burn exploded into the Lower North Fork Fire that destroyed the Kuehster Road neighborhood and killed three of its residents a week later, Colorado's state climatologist fretted that historic heat and drought were priming the state for an epic fire season. He was just as concerned about how he could get people to listen if he talked about it.

"I get, every year, the opportunity to talk to hundreds, probably thousands, of people face-to-face," he told me when I visited him in his office at Colorado State University's Department of Atmospheric Science. "They will see what I showed them last year with one more data point added on, which is how we did last year."

Although his job title had him saying "climate" thousands of times, he rarely combined it with the word "change." Not one of the 170 slides he had shown at the Colorado Farm Show that year discussed global warming.

Nolan didn't deny that Colorado and most of the West were warming and drying, but he stuck to the data. Coming to a conclusion about what it showed could alienate some of his most important audiences.

"You see change," he told one reporter. "I see variability."

Nolan is over six feet tall, with graying hair, a caterpillar mustache, wire-rimmed glasses, and a horsey grin. His enthusiasm for data can squeeze guffaws out of him as he looks at the driest of graphs. He's the quintessential weather geek.

Despite the huge and deadly wildfires of March, Nolan noticed that most of Colorado was just enjoying the early end to winter. "People were eating outdoors at all of the restaurants up and down the Front Range," he said. "Even in the evenings. Even in the ski towns. People were saying, 'This isn't right, but it's nice.'"

But the weather was making Nolan increasingly uncomfortable. "We were running 10 to 15 degrees warmer than average, which meant that we were running a full five to six weeks ahead of average in temperatures," he said. "May temperatures in late March."

Spring is when Colorado gets most of its moisture, but the high temperatures of 2012 didn't come with any precipitation. Grasses that sprouted weeks early and trees that leafed out in winter sucked up what water was left in the ground. "The pleasantness was masking this incredibly rapid dry-out," Nolan said.

At the farm where he and his wife raise horses and chickens, two inches of dust covered corrals that in a normal March would have been thick with mud. Dust storms the state had eliminated with soil conservation programs decades earlier came back in 2012 and were beyond anything Nolan had ever seen. Walls of soil hundreds of feet tall blew across the plains. Clouds of smoke sometimes followed.

"If it gets dry enough, it doesn't matter what conservation practices we've had," he told me.

The Lower North Fork Fire was another harbinger. "It burned right through snow," he said.

Still, he held out hope that a few big, soaking storms could bring back the water supplies and stop the fires.

But the mountains turned brown rather than white. Peaks barren of snow late in the season can't regain their white cover, even if they get a heavy snowfall. The albedo effect, in which dark ground absorbs heat from the sun that the white surface of the snow would reflect, warmed the mountains, ensuring that any new snow just melted off.

And the sky wasn't even trying to catch up.

"Things fell apart in April and just went *psshhh* down the tubes," Nolan said.

Nolan saw that the conditions in 2012 looked much like they had in 2002, when the massive Hayman Fire climaxed what was, until then, Colorado's worst fire season. Except that 2012 had warmed and dried out a month earlier.

IN JUNE NOLAN FINALLY GOT a taste of wet, green spring during a vacation to Michigan. Back in Colorado, he saw his worst fears manifest themselves.

Ten years to the day after the Hayman Fire ran more than 16 miles in 24 hours, Nate and Cindi Johnson were camping with their children on Crystal Mountain when they saw a puff of smoke where lightning had struck a ridgeline about 15 miles west of Fort Collins.[1] A firefighter with the Rist Canyon Volunteer Fire Department — a donation-funded operation — tracked down the smoke on his ATV. Shortly after 9 a.m. a single engine air tanker (SEAT) began dropping retardant around the blaze. By the afternoon, the U.S. Forest Service, Colorado State Forest Service, Poudre Canyon Fire Protection District, and Poudre Fire Authority had joined the fight. But by then the blaze was unstoppable. "We watched 300-foot flames from 80-foot trees," Nate told a Denver television reporter.

The High Park Fire exploded the same distance, almost to the mile, from my home in Boulder as the Lower North Fork Fire had. But this time the thunderhead of smoke spreading from the mountains across the plains was due north rather than south. And unlike the Lower North Fork Fire, which firefighters contained in a few days, the High Park Fire would run wild for nearly a month.

By the time I got to the growing cluster of satellite trucks and reporters along the Cache la Poudre River, the sheriff had already evacuated half a dozen neighborhoods, and he would be evacuating more for the next three weeks. Rist Canyon volunteers managed to save the historic, one-room Stove Prairie Elementary School while one of their own homes burned.[2]

Two calls to evacuate went to Linda Steadman, a grandmother

staying at the family cabin on Old Flowers Road, but both ended up in her voicemail. When a deputy and firefighter went to her property, a locked gate and the approaching fire turned them back. Two days later Larimer County sheriff Justin Smith confirmed her death.

"Linda Steadman, mother, grandmother, sister and wife, perished in the cabin she loved," her family said in a statement.

The sheriff reported that Steadman's cabin was just one of more than 100 structures that might have burned. He'd heard "very bad reports up in the Rist Canyon area."

Evacuees moved into motels, stayed with friends or family, or pitched tents at local campgrounds. By Monday, its third day burning, the fire had grown to 41,140 acres, and a Type 1 Incident Management Team — the highest level of disaster response in the nation — took charge.

"The hope for containment today is tenuous — totally dependent on the weather," Bill Hahnenberg, the team's commander, reported. "We may be at 0 percent tonight."

With the weather remaining unseasonably hot, dry, and windy Hahnenberg cautioned that the fire would "grow dramatically" despite five heavy air tankers, five SEATs, and five helicopters bombing the fire from the air, along with 500 firefighters from around the country engaging it on the ground. Crews took rafts across the Cache la Poudre River, which ran black with runoff from the fire, only to retreat when the blaze nearly overran them. Twenty-foot flames raced at speeds up to 40 feet per minute.

"It's a very aggressive fire. Fuel driven, wind driven, and the winds have not been favorable," Nick Christensen, Larimer County executive officer, reported. "The brush, timber, and grasses are very, very dry."

EARLIER THAT DAY the journal *Ecosphere* added fuel to a different fire when it published a study showing that changing precipitation patterns and increasing temperatures brought on by global warming would result in an increase in wildfires across more than 60 percent of the earth's land by the end of the century.[3] Nearly 40 percent would see more fires by 2050. Mid- to high latitudes, including the American West, would see the greatest increase in wildfires. One map indicated in red areas where climate models agreed there would be an increase in wildfires by 2100. Russia, China, Canada, and the United States looked like they were smeared with blood. Northern Colorado, where the High Park Fire was burning when the study was released, was the color of a fire truck.

"We have good records from across the United States, both thermometers and satellite observations, and they all come together and they tell us that 2012 was the warmest year that the lower 48 has had since we started taking records," Jim White, director of the Institute of Arctic and Alpine Research at the University of Colorado, told me.

June 2012, the warmest June on record in Colorado, had temperatures 6.4 degrees above average. "The conditions we saw in 2012 will be an average year in 2030," Nolan Doesken told me. "And a hotter Colorado is a more-vulnerable-to-wildfire Colorado."

KATHARINE HAYHOE, director of the Climate Science Center at Texas Tech University and one of the authors of the paper in *Ecosphere*, was accustomed to fiery landscapes, both literally and politically.

Like Nolan Doesken, she had close encounters with wildfires while doing her research. The year before her paper's publication, Texas had 28,000 wildfires — more than twice what the state normally endures. The blazes included the Bastrop Fire, which destroyed nearly 1,700 homes, making it the most destructive fire in state history. It was one of nearly a dozen fires that destroyed homes that Labor Day. Another was at Possum Kingdom Lake, which burned twice that year. Katharine has a home there.

"As a series of wildfires swept through central Texas lake country, first in April, then again in August, we went to bed night after night convinced that our place would be gone by morning," Hayhoe wrote on the blog *God's Politics* a few days after her paper was published. "With 100-ft. flames, near-zero humidity and 50 mph winds, all I can say is that we were lucky. Many of our neighbors, some a few hundred yards away, were not."[4]

The fire stopped 200 yards from her house. One of her neighbors lost her home to the first Possum Kingdom wildfire in April, bought a new house, and lost it to the August fire.

"The fires were really unusual in terms of their speed and their ferocity and their intensity," Hayhoe told me. "You barely had time to get out of the way."

Hayhoe held vigil from afar, staying up all night at her computer and sharing bits of news with neighbors. "When you have a personal connection, it just makes all the difference," she said as she described photos of the fire. "You're trying to figure out, 'Is that your place?' That burned skeleton with a washing machine sitting there between these burned beams?"

So when the paper came out in *Ecosphere,* Hayhoe was happy to push it into the public sphere. "When you talk to people about climate change, you have to be prepared to tell them why you care," she told me. "And it better not be some kind of airy, fairy reason. For me this was one more reason why I personally care. I can say, 'Look, here's the fire . . . Here's what it did to our neighbors.'"

Hayhoe knew she was in for more heat. She didn't just share Nolan's familiarity with wildfire threatening the family home; she'd also experienced the consequences of reporting how human impacts on the climate are driving wildfires. Six months before the publication of her paper, presidential candidate Newt Gingrich dropped her chapter about climate change from his book about environmental entrepreneurship after conservative firebrand Rush Limbaugh lambasted the work of "the climate babe" on his radio program.

"That was not the first time it had happened," she told me. "I was on the O'Reilly [Fox News] show a couple years ago. I got over 200 hate mails the very next morning, and who knows how many blogs."

Hayhoe had published for years in scientific journals and never received a hateful email or been attacked in a blog. But that changed when she spoke of her research on climate and wildfire. "I've gone from zero to a constant background [of attacks]," she told me. "Every time you come out in some public way that climate change is real, the attacks peak, and they don't go back down to where they started from. They keep at a bit of a higher level."

In her office in Lubbock, her finger traced the air like she was drawing foothills rising jaggedly into a mountain range. The air graph she drew of the political heat coming down on her mimicked graphs that I had in front of me showing the increase in temperatures and wildfires in the West.

"So the next time it peaks higher, and then it drops back down to this level," she said. "And the next time it peaks higher . . . There's a couple of bloggers, I swear they probably blog about me on a weekly basis — people who seem to be fairly obsessed in ways that are a little bit . . . unhealthy."

But Katharine Hayhoe's unusual mix of credentials makes her a difficult target for deniers of climate change. She holds a PhD in atmospheric science from the University of Illinois, has received scores of fellowships and grants, and was a reviewer for the Nobel Prize–winning Intergovernmental Panel on Climate Change (IPCC). But she is also an evangelical Christian and is married to a pastor.

"I'm happy to stand up and say I'm a Christian. I attend a Bible church," she told a reporter from *Mother Jones*. "In the past they've really tried to paint scientists as godless liberal treehuggers."[5]

To counter that impression, Hayhoe and her husband, Andrew Farley, wrote the book *A Climate for Change: Global Warming Facts for Faith-Based Decisions*. Inspiring Christians to care about the climate requires separating faith and politics, she said. "If you control for politics, then most of the religious bias falls away. People have been allowing their political party to inform their statement of faith . . . It's almost gotten to the point where the church has a Republican statement of faith."

Christians who vehemently deny climate change aren't getting that message from the Bible. "Where they're getting it from is, more frequently than not, conservative media, which has this thin veneer of Christianity on it, but it is not Christian in any way, shape, or form . . . If it was convenient for them to call themselves a Buddhist, they would."

But, at least in Texas, once a bastion of climate change denial, the fires appear to have burned away some of the skepticism. Two years after the fires, a Yale University survey showed that 7 out of 10 Texans believe the climate is changing.[6] That doesn't mean they accept any responsibility for the droughts, heat waves, and fires, however. Only 4 in 10 respondents said they would ascribe the changing climate to any kind of human activity.

"There's been a major progression in people's willingness to admit that there is a problem, but we've seen zero progression in the acknowledgment of humans as a cause of the problem," Hayhoe said. "In 9 out of 10 cases I'd be willing to bet that people object to the cause of climate change because they object to the solution."

But even among the self-interested, wildfires prove a powerful persuader. After the Bastrop Fire lawyers for the electric utility that owned the power line blamed for starting the blaze contacted Hayhoe. They wanted to know if they could attribute the fire to global, or at least to Texas, warming.

She sees attributing the increase in fires to climate change as more convincing to business and political leaders, and to the public, than attributing it to other impacts that will likely be more damaging to society but are less dramatic. "The problem with climate change is that it's been a distant, far-off problem. It's happening in the Arctic, it happens with ocean acidification, it happens with slow sea-level rise," she said. "Fire, that's immediate and catastrophic."

ACCORDING TO THE ECOSPHERE PAPER, even some parts of the planet predicted to experience an increase in rain will see an increase in wildfires, too, as intense "precipitation pulses" feed bursts of vegetation that will burn when drier conditions return. In other areas, the converse will likely prevail, as occasional deep droughts will bring fire to areas that are otherwise consistently moist.

When I spoke to Hayhoe, she'd just finished collaborating on another paper that was published in April 2014.[7] It modeled precipitation patterns in North America during the next century. Alaska, she noted, is projected to get wetter while the Southwest gets drier. "But . . . even though it's getting wetter in Alaska, it's getting more drought-prone. And even though it's getting drier in the Southwest, it's getting more prone toward heavy downpours at the same time as it's getting drier."

Both those scenarios could lead to more fire on the ground.

FOREST FIREFIGHTERS LEARN THE "FIRE TRIANGLE": heat, oxygen, and fuel. The *Ecosphere* paper cites a different triad driving the increase in wildfires: resources to burn, atmospheric conditions, and ignitions. The changing climate will impact all three sides of that triangle.

In the first category, a warming climate will change the amount and types of plants that grow, and burn, in any given area.

On the second side of the triangle, annual variations in the atmosphere, such as droughts and heat waves, influence how flammable a landscape is, while short-term changes in fire weather — the heat, humidity, and winds of a given day — determine whether a flame will take off or just smolder.

Finally, there are ignition sources, the only natural one of note being lightning. Research led by David Romps, an atmospheric scientist at the University of California, Berkeley, predicts that for every degree Fahrenheit the earth's atmosphere warms, there could be nearly 7 percent more lightning strikes.[8] Current predictions of a seven-degree increase in temperature by the end of the century would mean as many as 50 percent more bolts of electricity hitting the ground.

And in a hotter climate, even human ignitions will likely increase. Power lines, for example, are more likely to arc and spark when overtaxed by the increased use of air conditioners in hot weather.

Even areas predicted to see less fire may suffer negative consequences, Hayhoe said. "Increases and decreases can both be bad for the natural ecosystem, because wildfire is part of what makes it healthy."

Regardless of how meticulous her research is, she knows that her time spent in the public eye and in her church is more likely to bring change. "I don't think throwing more scientific facts at the problem is going to solve it," she said. "Facts don't win the argument. We've had enough facts on climate change for decades already."

NONETHELESS, THE FACTS tying the warming climate to a more fiery planet keep coming. Six months after the *Ecosphere* paper was published, a study by Harvard's School of Engineering and Applied Sciences predicted that the fire season in the western United States could grow nearly three weeks longer by 2050.[9] That is on top of the 78 days that the average wildfire season in the West has already increased, according to a study of the years 1970 through 2003. The fires that ignite will produce twice as much smoke, the Harvard study reported, and burn substantially larger areas. In a 2015 paper in the journal

Nature Communications, Matt Jolly, an ecologist at the Missoula Fire Sciences Laboratory, reported that from 1979 through 2013 wildfire seasons expanded almost 20 percent, and the amount of land vulnerable to fire nearly doubled.[10]

"It turns out that, for the western United States, the biggest driver for fires in the future is temperature," said Loretta J. Mickley, an atmospheric chemist at Harvard and coauthor of the study there. "When you get a large temperature increase over time, as we are seeing, and little change in rainfall, fires will increase in size."[11]

The Harvard paper predicted that the area burned by wildfires in the Front Range of Colorado will increase by between 70 and 100 percent. The Southwest could see as much as a 150 percent increase in area burned.

In the West warmer temperatures' impacts on wildfire are most obvious in the snowpack. As temperatures warm, the snowpack dwindles and melts off earlier in the year.

"The snowpack really kept a lid on the fire danger," Steve Running, one of the Nobel Prize–winning IPCC authors and Regents Professor of Ecology at the University of Montana, told me. The earlier melting of mountain snowcaps is largely responsible for the extended fire season in the mountains of the West. He added that the reduced snowpack has extended not only the length of the fire season but also the range of flammable vegetation. Species that snow used to hold back from the high country are now encroaching into it.

"Especially in the West, the dynamics that are causing increasing fire frequency are lower winter snowpack and an extension of the dry season," Running said. "When the snowpack melts off . . . combustible fuels move up the mountain."

The greenhouse effect also make fires like the High Park Fire burn intensely through the night — a phenomenon many of the firefighters reported they hadn't seen before. Typically, cooler temperatures at night help vegetation regain some of its moisture. If temperatures fall far enough for dew to cover the ground, that dampness will slow fires. But greenhouse gases trap the daytime heat, keeping nights sultry and dry, so fires keep burning in the dark.

· · ·

Alaska had its hottest may in 91 years in 2015. The next month 1.8 million acres burned there in just 12 days, nearly twice the previous record for acres burned over *an entire June*. Some 320 fires burned across the state, charring nearly half a million acres in a single day.[12]

In 2015 Climate Central reported that Alaska was warming almost twice as fast as the rest of the nation due to the fact that polar regions were heating faster than the rest of the planet. At the same time, the average annual acreage burned in the state had doubled each decade between 1980 and 2009.[13]

While the increasing acreage in flames was scary, what was actually burning was just as frightening, if less dramatic. In addition to vast evergreen forests, the fires burned deep into tundra and permafrost, which hold about twice the amount of carbon as is already in the atmosphere, as well as huge stores of methane, an even more powerful greenhouse gas. One study showed that 60 percent of the climate-warming gases released in a large Alaska fire came not from burning trees and vegetation, but from the combustion of organic material in the soil.[14]

With the wildfires themselves adding huge amounts of greenhouse gases to the atmosphere and thus warming the climate, they'll likely drive even more fires.

"Fire has a substantial positive feedback on the climate system," a 2009 paper in the journal *Science* reported of the phenomenon in which fires heat up the atmosphere.[15]

When nolan doesken and his wife returned to their house after their vacation to Michigan, they found ash and cinders from the High Park Fire on top of the dust they were already cursing. To get to his office, on a hill between Fort Collins and the national forest, he had to show his credentials to armed national guardsmen at a checkpoint. He passed the incident command center where Colorado senators Mark Udall and Michael Bennet addressed the press. His office window looked out on the massive plume of smoke, and he sometimes saw flames leaping above the Horsetooth Reservoir.

The deepest burn to his psyche, however, came not from the flames, but from the water in his irrigation ditch. Ash had clogged a neighboring family's pump, which had shorted out and electrocuted their young daughter.

"Our daughter could have been doing the same thing just a few years earlier," he told me. "We both irrigate out of the same irrigation ditch . . . It just made me ache."

Nolan knew he'd be seeing impacts trickle down from the fire for years. Fish, farmers, factories, and families all shared the water supply filling with ash, cinders, chemicals, and debris from the fire. "It's amazing watching the cascading effects," he told me. "The hydrologic consequences, the agricultural consequences."

But Nolan was most worried about the changing climate, which would bring even more conflagrations, along with the debris flows and floods that follow. He began thinking about how he could get Colorado to listen when he finally began talking about those threats.

18

FIREBUGS

Rocky Mountain National Park — June 23, 2012

BY THE THIRD WEEK OF JUNE, the High Park Fire had destroyed 191 homes, making it the most destructive blaze in Colorado history. It would hold that title for two weeks.

I drove steep and winding dirt roads to access neighborhoods tucked deep into the mountains, and stood at roadblocks and outside shelters with residents who wondered when they would be allowed back into their homes — or if they had homes to return to. Some of the thousands of evacuees wouldn't go back, even if their houses had survived. After one meeting of evacuees, Ellen Bozell told me that she, her husband, and their two children were living in a friend's barn. "If our home burns, we won't rebuild up there," she said.

Some 50 guests were evacuated from the Shambhala Mountain Center, a Buddhist retreat in the mountains southeast of the Cache la Poudre River. When the fire jumped the river, 10 volunteers stayed behind to cut trees and mow grass to protect the Great Stupa of Dharmakaya — one of the largest Buddhist shrines in North America. They left the retreat's cabins to burn.[1]

Workers at another sanctuary in Rist Canyon shot tranquilizer darts into the wolf dogs there and then moved 17 of the sedated animals to a kennel away from the fire. But before they could round up the rest of the animals, flames drove the caretakers away. Fire dens dug four feet into the mountainside were the only shelters for the remaining wolves. Bunkers had never been used to shelter wild animals from a wildfire before, caretaker Michelle Proulx said, and she had no idea if the animals would survive.

A worker allowed back into the sanctuary after the fire had passed found charred ground around the dens.[2] "You see scorched trees, scorched ground on the left, untouched ground on the right," he said in a video he shot there.

Then bushy gray fur moved through the burnt trees, "and a wolf, happy and healthy," appeared.

At Paradise Park a donkey named Ellie, her whiskers singed, walked up to firefighters in a pasture. Behind her were three horses and another donkey, their tails curled from the heat, that she had led to safety.[3]

Temperature records were falling almost every day. Beginning Friday, June 22, Denver recorded temperatures above 100 degrees for five consecutive days. Two days, at 105 degrees, would break the all-time record temperature for the city.[4] On Saturday a friend and I headed to the high country of Rocky Mountain National Park, where the air would be a little cooler, and I could do a bit of rock climbing and take a breather from the smoke.

Just before I started climbing on Lumpy Ridge, I heard a two-way radio a few feet away, then saw the uniforms of two National Park Service climbing rangers, Adam Baxter and Jess Asmussen. It turned out Adam and I were both planning to attend the wedding of a firefighting friend in a few months. We were laughing at how the brotherhood of the rope and the fellowship of fire are tight-knit communities, with occasionally interwoven strands, when their radio interrupted us.

"It's crowning," it squawked. "It's in the trees. Houses threatened."

Beyond the forested ridgeline to our south, we could see smoke rising from Estes Park. "It sounds like it's on the ground," Jess said, "right off the road."

A few hundred feet from the Beaver Meadows gate to the national park, high winds had rubbed a cabin's power line against a tree. The wire had frayed and ignited the bark. Gusts had carried the flames into tight stands of pines intermixed with clusters of homes.

Jess's cellphone rang with a call from his wife. "Her best friend just showed up," he said after hanging up. "The fire broke out across the street from her house. She grabbed an armload of stuff and got out. She's moving into our place."

The Woodland Heights Fire burned 21 homes that afternoon.

Two choppers and an air tanker came from the High Park Fire, but that blaze was also blowing up. Resources were being spread thinner by the hour.

That night Larimer County sheriff Justin Smith choked back tears while addressing the press. Both blazes were under his jurisdiction, and both had destroyed homes.

The High Park Fire's latest run had started on Thursday, and by Saturday night it had grown from 50,000 acres to 81,000, making it the second-largest fire in Colorado history. It was already the state's most destructive in terms of property when it roared up the narrows on the Cache la Poudre River to burn at least 10 more homes Friday and force another 1,700 evacuations. Smith said they knew more homes were burning, but firefighters were too busy to count them.

On Friday Colorado governor John Hickenlooper signed an executive order releasing $6.2 million in emergency funds to fight the High Park Fire along with three others across the state.[5]

Then, on Saturday, eight new fires broke out, including the one I witnessed from Rocky Mountain National Park. There was more terrain burning at one time than ever before in state history. Overwhelmed firefighters could give little attention to fires that normally would have been top priorities. Fire commands were competing for resources, Bill Hahnenberg, the incident commander for the High Park Fire, told reporters.

At 3:45 that afternoon, around the time that firefighters were getting the Woodland Heights Fire under control, the State Line Fire ignited just south of Durango, threatening oil and gas pads, along with homes that were "inaccessible to firefighters."[6] In central Colorado the Treasure Fire ignited near Leadville, raining ash on the resort community of Breckenridge.[7] The Weber Fire, an adolescent's arson on the Western Slope, burned 2,500 acres of scrub oak killed by a frost that followed the early spring. That fire was suspected of igniting seven smoldering coal fires that were discovered the following year in piles of refuse coal, mines, and coal seams.[8]

But most concerning was the Waldo Canyon Fire, just west of Colorado Springs. Over the previous weeks firefighters there had responded to a series of 25 suspicious blazes. Most of them were clearly "human starts" but were easy to extinguish. Smoke from the Waldo Canyon Fire, the latest in the series, was first reported near a trail in its namesake canyon at 7:50 Friday night. During the next two hours firefighters from the U.S. Forest Service and other agencies hiked trails near the smoke, but not the Waldo Canyon Trail, which would have led them to the blaze. Early Saturday a jogger encountered a smoldering fire along the trail. He called authorities, but the dispatcher he spoke with failed to record his location, name, or phone number, or to pass his information along to the El Paso County Sheriff's Office fire center. Despite increasing reports of smoke, firefighters didn't find the blaze until five hours later.[9]

They sent crews and a single engine air tanker, but when the temperature and winds rose, the fire took off fast. At 3:12 p.m. the city ordered mandatory evacuations for more than 3,000 homes — less than 10 percent of the total they would evacuate in the coming days. That afternoon the City of Colorado Springs closed its iconic Garden of the Gods park for the first time since 1901. The fire burned downhill toward developments that night, traveling as much as half a mile in an hour and growing to nearly 2,000 acres in size. After midnight authorities went door-to-door to evacuate residents of Manitou Springs, a historic enclave of tourists and artists, in hopes of getting everybody out by dawn. They didn't finish the evacuation until about 9 a.m., when the fire was barely a quarter mile away.

The fire outside Colorado Springs, one of 16 large fires burning out of control across the country, was the top firefighting priority in the United States almost as soon as it ignited. Between all of Colorado's fires, the $6 million in emergency firefighting funds the governor released Friday was all but gone by the time the weekend was over.

CLIMATE, HOUSES, POWER LINES, OIL PADS, and even marijuana operations and meth labs hidden in the woods all brought more wildfire to Colorado. But another change in the forest had also turned millions of acres of pines red, though not with flames.

Since the late 1990s an outbreak of the mountain pine beetle, a pest the size of a match head, had devastated more than 107,000 square miles of forest in the United States and Canada—an area larger than the entire state of Colorado. In British Columbia the pest killed more than half the commercially valuable pines in less than 20 years. The beetle is a native of the West, and it breaks out in regular cycles, but the outbreak that began in the late 1990s was 10 times larger than any other on record.

The beetles use a pheromone to call an army of insects to an individual tree and then burrow inside. They carry a fungus in their mouths that, along with the beetles' tunnels, blocks the tree's circulatory system and kills it. Trees can push the beetles out with resin, but when weakened by drought, they can't produce enough sap to defend themselves. As they die, their needles turn red. Entire mountainsides turn a rusty crimson, then transform into gray "ghost forests" after the needles fall off.

While it was the great lodgepole pine forests of the Rocky Mountains that the pest devoured in the most recent outbreak, it also infests many other evergreens. The effects of the tiny insects' destruction cascade through ecosystems. The beetles have so devastated whitebark pine forests around Yellowstone that at the end of 2011, the Ninth Circuit Court of Appeals put grizzly bears, for which the whitebark's cones are a critical food source, back on the endangered species list just four years after the U.S. Fish and Wildlife Service had taken them off.[10]

In 2012, as the march of the pine beetle slowed (the insects were running out of mature lodgepole pines to eat), the spruce beetle infested a million acres of Colorado's high-elevation spruce forests.

As the High Park Fire burned into forests of beetle-killed timber, the debate about the impact of those dead and dying trees on fire behavior also grew heated. By 2012 at least 39 separate studies had looked at how the millions of beetle-killed trees influenced the behavior of wildfires, but there was no clear picture of how they would burn.[11] I hiked over mountains covered in trees that seemed color-coded for fire, which was how the best-known early hypothesis saw them: When the trees were in the red phase—dead but still holding rust-colored needles—they would be prone to severe crown fires that would burn hotter and faster than in a green forest. But in the gray phase, when all the needles had fallen, fires would have difficulty moving through the crowns and calm down.

A study led by Matt Jolly at the U.S. Forest Service's Rocky Mountain Research Station showed that the red, dead needles ignited three times faster than green ones.[12] Crown fires in red forests, some firefighters and foresters theorized, would ignite with lower wind speeds and less-dry conditions than in forests unaffected by the beetles. The intensity of the crown fires in beetle-killed forests, other researchers predicted, could launch embers farther, thus spreading fires faster.

Other models showed that dead trees that fall to the ground, along with the increased wind penetration in stands with fewer trees and needles in the canopy, would increase the intensity of ground fires. The hotter fires on the forest floor would drive crown fires to erupt with less wind than they typically require. Researchers also documented beetle kills increasing the temperature on forest floors as more sunlight shined through a thinned canopy.

Some studies confirmed that gray forests with no needles on the trees actually slowed down crown fires.

But four months before the High Park Fire ignited above Fort Collins, a report from the Joint Fire Science Program in Boise, Idaho, argued that neither red nor gray forests would be likely to burn more severely than green forests, largely because the death of the trees reduced the amount of fuel in the canopy.[13] The paper also maintained that climate and weather factors, rather than needles and dead trees, drive most wildfires.

In fact, during the three biggest fire years since 2000, including the 2012 season, research showed that fires didn't burn more in forests infested with mountain pine beetles than in those unaffected by the insects.

Lodgepole pines tend to die in "stand-replacing" events. Wildfires or bugs kill off entire forests every 100 to 300 years. The trees' cones, sealed tight by resin, release their seeds in response to the heat of a fire, and seedlings flourish in the carbon-rich soils after a burn. So these trees have evolved to encourage fire in order to reproduce. During a hot drought like Colorado's in 2012, lodgepole forests are prone to burn big, regardless of whether they're green or red.

In the High Park Fire, however, many firefighters reported seeing crown fires that doubled or tripled their speed and intensity when they came to red, beetle-killed lodgepole pine stands, and they couldn't care less what scientific models said the fires would do. Kevin Moriarty, a graduate student at Colorado State University, interviewed firefighters on blazes between 2010 and 2012 in northern Colorado. They described fires that launched more embers to start spot fires; ground fires that transitioned much faster into crown fires or did so without the ladder fuels that are usually required to take a fire from the forest floor to the canopy; and fires that raced through red-needled treetops far faster than green ones.

"I saw fire running in spruce beetle kill like it was grass or sagebrush," Fred Schoeffler, a firefighter from southern Colorado, told me at the Large Fire Conference in Missoula, Montana, where Moriarty described the results of his interviews in 2014.

Nobody disputed, however, that the two things turning western forests red — beetles and wildfires — have a common parent: the climate.

At the University of Colorado's Biogeography Lab, just a few minutes from my office at the university, I visited Tom Veblen, who for 25 years has studied how insects and fires impact forests around the world. Tom coauthored several of the papers that found little evidence of mountain pine beetles increasing the severity of wildfires, and the lab's studies of fires in beetle kill during the drought of 2002 showed that they didn't behave much differently than they did in green forests. He wasn't, however, quick to dismiss what firefighters and land managers have observed. The science surrounding wildfires, he pointed out, is tangled — although some aspects are quite simple.

"There's one thing that's consistent where we're seeing the increase in wildfire and mountain pine beetle," he told me. "It's warmer."

The only thing known to stop a mountain pine beetle outbreak is prolonged subzero temperatures, which haven't occurred in a decade in much of the Rockies and western Canada. Climate models predict that those temperatures will become increasingly rare. Trees stressed by drought and heat waves are more susceptible to pests, and the insects thrive in the warmer temperatures. The University of Colorado's Jeffry Mitton and Scott M. Ferrenberg found that due to the warming climate, mountain pine beetles in some Colorado forests have started reproducing twice a year, instead of once, leading to an exponential increase in the number of hungry bugs.[14]

In the end, the greatest risk to firefighters may not be in the red or gray forests today, but decades from now after the beetle-killed trees have fallen to the forest floor and new canopies of pines grow above them. Fires there would be difficult to fight because heavy timber burning on the forest floor would hinder firefighters' travel through the forest and push crown fires to burn faster and hotter.

WHILE THE DEBATE RAGED about whether beetles had made the High Park Fire hotter and faster, a blaze ignited on the Colorado plains with the help of another pest. But though the Last Chance Fire quickly turned into the fourth-largest conflagration in Colorado history, flames erupting all over the state made the grass fire as easy to ignore as the community it was named for and the invasive grass that helped drive it.

The town of Last Chance was born of the automobile, and killed by it more than once. It was the last chance to gas up on U.S. 36 when that was the main thoroughfare between Kansas and Denver. Thousands of pickups, cattle haulers, and tractor-trailers stopped at the pumps there. But when the construction of Interstate 70 rerouted that steady stream of traffic in the 1960s, it starved the town.

On June 25, 2012 — the day the fires ignited from Rocky Mountain National Park to Colorado Springs — Last Chance, population eight, was a graveyard of abandoned, decaying motels and a burned-down service station. Only the recently closed Dairy King, with its custard-yellow walls, showed any of its former flair.

But fuel and the automobile once again put the town on the map. A car traveling on State Highway 71 blew out a tire, and sparks from the rim showered dry grass in a ditch five miles south of town. In just over a day the subsequent blaze burned some 45,000 acres and destroyed 11 structures.

Firefighters blamed triple-digit temperatures, the drought, and high winds for the blaze's ferocity. Ecologists and ranchers in the area noted another factor. Unlike the massive red forests of beetle-killed trees that are impossible to miss in the mountains, the invasive pest bringing fire to the plains is barely noticeable, even when it spreads out over thousands of acres. It's just a stalk of grass, but there's little debate about its relationship with fire.

Since the 1800s cheatgrass, a wispy brome native to Asia, has spread around the world, usually hitching rides in packing materials and grain exports. By 1900 expanding railways had dropped the grass in Washington, Utah, Wyoming, and Colorado.

The weed can germinate at any time of year but usually chooses the fall. While it's covered in snow during the winter, its roots continue to grow, which allows it to sprout early and "cheat" native vegetation out of moisture and nutrients. It can reach 20 inches tall while native plants are just getting started. Merely two inches of precipitation can drive a strong flush.

But the invasive weed uses some even sneakier tricks to drive out native vegetation. Cheatgrass completes its life cycle quickly, building a prolific seed bank, then drying out while native vegetation is still young. A bloom of cheatgrass can leave nearly 300 pounds of dry fuel per acre for hungry fires.

Bunchgrasses native to the high plains and the Great Basin evolved to hinder the spread of fire. They grow in clumps separated by bare soil and sand. It's difficult for the flames on one bunch to reach others.

But cheatgrass grows in "continuous fuel beds" that are so thick and even they are often mistaken for Kentucky bluegrass or even fields of wheat. With no natural fuel breaks in the tall, dry grass, fire spreads easily. A 2013 study reported that landscapes infested with cheatgrass are twice as likely to burn as those covered with only native grasses.[15]

"Cheatgrass was introduced well over 100 years ago, but it . . . didn't take hold in a very significant way as far as carrying [fire] until about 30 years ago," Tim Murphy, the assistant director for fire and aviation at the Bureau of Land Management, told me when I visited the National Interagency Fire Center in Boise — America's Pentagon of wildland fire.

The BLM manages the most land of any federal agency, mostly wild grasslands and scrub, the landscapes where cheatgrass has the most impact. With a neatly trimmed, graying beard and an easy smile, Murphy seemed cheerful as he described the cheatgrass flooding Idaho, Nevada, California, Utah, and, increasingly, Wyoming and Colorado.

"It used to take one hell of a wind to carry a fire," he told me of the sagebrush steppes where cheatgrass is, literally, spreading like wildfire. "A 30,000-acre fire, . . . you would write home about. Now a 300,000-acre fire is the stepping-off point for comment. Our whole view has changed."

The blazing cheatgrass kills off the native plants before they mature, preventing them from regenerating and spreading its own abundant seeds. Native vegetation that survives a fire often struggles in the scorched ground — some arid and treeless landscapes take 70 years to recover after they burn. Cheatgrass, however, snaps back quickly.

Fires in native grass landscapes rarely burn more often than every 25 years, and sometimes as infrequently as every 70 years. "It's down to 10 and even five years in some places" where cheatgrass is present, Murphy told me.

So it's a vicious, and fiery, cycle — more cheatgrass brings more frequent and larger blazes, and the increase in fires brings more cheatgrass.

The invasive grass has one more advantage, which Tim illustrated with a map of fires along Interstate 84 around Boise. Spreading out from the highway and housing developments was a growing red blotch showing fires. As the weed spreads over the landscape, the sparks people produce where they live, play, and travel — like the Last Chance highway in Colorado — are far more likely to start fires.

Today cheatgrass covers more than 101 million acres of the United States. In 1999 cheatgrass-driven fires burned more than a million acres of the Great Basin in 10 days. The pest infests about 46 million acres of winter wheat annually, which costs growers about $300 million in lost crops and an additional $70 million in herbicides to try to kill it off.[16] It dries up too early to be of much value for grazing livestock. And it devastates the habitat of many other animals. The sage grouse is teetering on the brink of threatened species status, in part due to cheatgrass fires burning away its nesting sites in sagebrush.

Bethany Bradley planned to search for extraterrestrials on Mars, but ended up studying aliens on earth. The biogeographer at the University of Massachusetts is one of the leading researchers studying the link between cheatgrass and fire in the American West.

Little green men seem more exciting than a dull green grass, but

Bradley found a way to get her outer space fix using satellite data to map cheatgrass distribution. "We didn't see any trends in vegetation change, but we did see this really weird signal on parts of the West that greened up really strongly. They basically looked like a bunch of little golf courses in 1998," she recalled, noting that a year earlier they looked like the rest of the desert.

The El Niño of 1998, she realized, brought a change in precipitation to the region. Year-to-year variations in precipitation favor cheatgrass. "Most of the desert-adapted species can't do anything with all of the extra water," she explained. "But cheatgrass can."

In the long run, however, cheatgrass thirsts for only one thing. "It's really the fire cycle that triggers the biggest impacts on native species, native plant diversity, and native animal diversity," she said. Cheatgrass fires create ecosystems that are "monotypic, or single species as far as the eye can see."

The warming and drying climate of much of the West, Bethany said, is likely to increase the speed and spread of cheatgrass. Changing temperature and precipitation patterns are spreading the weed into once-unsuitable habitats in northern Idaho, Montana, and Wyoming.

By overlaying NASA satellite measurements of burnt areas across the United States with year-to-year changes in vegetation coverage, Bethany and her collaborators confirmed that cheatgrass is driving more frequent, faster-spreading, and longer-lasting fires in shrub-dominated landscapes. "The numbers showed it's basically burning twice as frequently as you would expect relative to its land area," she said. "One of the things that really surprised me a lot was the influence of cheatgrass on ignition of these fires."

Free-flying truck chains and sparks from power equipment ignite blazes more easily in cheatgrass than in other grasses. In addition, human interaction with infested land is increasing. "The fires aren't just burning all this land that nobody cares about," said Bradley. "They're actually burning the land that we live on."

Cheatgrass flourishes in disturbed terrain, and the epic floods and fires that plagued Colorado in 2012 and 2013 left the state primed for an epic invasion.

• • •

MARLYS SWAN LIVED 12 MILES NORTH of Last Chance's downtown but worked as a massage therapist in Brush, Colorado. She wasn't worried when she saw the cloud of smoke on the plains moving toward her house on the same day that I saw smoke rise in the high country of Rocky Mountain National Park. Grass fires are common in the flatlands, and local fire crews keep most of them from going very far. But by the time Marlys got to her brother and sister-in-law's house, near her own, they had received two reverse 911 calls. She raced home, grabbed her two dogs, and headed to the next town to the north. Five minutes after she got there, it also was evacuated. The fire was moving almost as fast as the evacuees. As she continued driving away from the smoke, her phone rang. "My brother . . . said, 'Your house is on fire,'" she recalled.

When she returned the next morning, only the chimney was standing. The rest—family photo albums, her beloved collection of books, her mother's wedding ring—were in a pile of ash and cinders that filled the basement.

Justin Wagers, in nearby Woodrow, Colorado, manages his family's grain fields to keep out cheatgrass. But in his second job, as a volunteer firefighter, there's no escaping it. He was in the middle of the wheat harvest when the Last Chance Fire broke out. He and his brother Jonathan were racing to it in their fire truck when they were suddenly blinded.

"The smoke came across the roadway, and we just didn't have visibility anymore," Justin said. "We went off into the steeper part of the ditch."

Flames engulfed their old fire engine, but the Wagers boys managed to escape.

When I heard about their truck getting burned over, I thought of the Struckmeyers, who had been burned over in their fire truck by the Heartstrong Fire about 75 miles to the west three months earlier. They were with the Wages Fire Department, so they were easy to confuse with the Wagers family, even without the remarkably similar circumstances.

Prairie fires typically stay close to the ground and burn so fast that they do little to hardwood trees and telephone poles. Not so in Last Chance. The grass fire burned down nearly 1,000 utility poles. "Grass that was maybe two to three inches tall was burning flames that were starting telephone poles on fire, 20 feet in the air," Justin said.

The fire jumped roads, and where it couldn't, it blasted through the culverts beneath them.

"I'd never seen anything like it. It was about 107 degrees that day. The humidity was two percent . . . perfect conditions for a fire," he recalled.

Part of those conditions, he said, was the invasive grass that draws fire like flowers draw bees. "I guarantee there was lots of cheatgrass that was burned," he said. "When you get a really wet fall and a really wet spring, you get a lot of cheatgrass that gets real green and nice, and then it's dead by the middle of June. After that it's nothing but a bunch of fuel for fire sitting out there."

Efforts to level the playing fields where the cheater now has an advantage include prescribed burns, herbicides, and fuel breaks. Researchers are even experimenting with absorbent polymers, like those found in disposable diapers, which can deprive the early-sprouting invasive of the moisture it needs to get a head start on native grasses. None have had much success in stopping cheatgrass's spread.

19

FOREST JIHAD

As the high park fire burned through beetle-killed forests, and invasive weeds sped the Last Chance Fire across the plains, a different kind of firebug threatened the forests above Colorado Springs and others around the world.

Just weeks before the Colorado fires, *Inspire,* the online magazine for Al Qaeda, charged jihadists to ignite wildfires as tools of terrorism.[1] The 11-page article titled "It Is Your Freedom to Ignite a Firebomb" noted that "in America, there are more houses built in the countryside than in the cities."

It cited the hundreds of fatalities and towns burned over in Australia as something terrorists could replicate, or even surpass. Instructions detailed how to construct "ember bombs" and timing devices, and described the forests and weather conditions in which they would burn most destructively.

"Fire is one of the soldiers of Allah," it announced.

The article wasn't the first to encourage using wildfire as a tool of terrorism.[2] "Summer has begun so do not forget the Forest Jihad," the Al-Ikhlas Islamic Network advised in a posting discovered by U.S. intelligence officials in 2008. It quoted imprisoned Al Qaeda terrorist Abu Musab al-Suri calling on "all Muslims in the United States, in Europe, in Russia, and in Australia to start forest fires."

"You can hardly begin to imagine the level of the fear that would take hold of people," the posting noted.

Alexander Bortnikov, chief of Russia's Federal Security Service, cited jihadist websites when he blamed terrorists for a series of fires in Europe that summer.[3] "This method allows al-Qaeda to inflict significant economic and moral damage without serious preliminary preparations, technical equipment, or significant expenses," Bortnikov said.

Captured terrorist Assem Hammoud, who was charged in 2006 with plotting to blow up train tunnels in New York, also admitted to planning arsons in California forests. Another Al Qaeda operative admitted to plotting wildfires in Colorado, Montana, Utah, and Wyoming, using timed devices that would ignite after the pyro-terrorists left the country.

Wildfire has long been a weapon. Native Americans and Australians used fire to push back white settlers. During World War II Japan launched more than 9,000 "fire balloons" over U.S. forests. They were largely ineffective, but in one case killed six Oregon picnickers who were attempting to drag a balloon out of the woods when it exploded.

While there are actually few substantiated cases of wildfires started by terrorists, arsonists started six of the nine fires studied in the United Nations' global assessment of megafires in 2011.[4] At the time of that report, the largest wildfire in Arizona history was 2002's Rodeo-Chediski Fire, which grew out of two arsons, one committed by an Apache looking for work on a fire crew, and the other by a woman signaling for help when her car ran out of gas. The report also listed Colorado's largest wildfire, the Hayman Fire, which burned during the same summer as the Arizona fire and turned into a legal drama.

Terry Lynn Barton, a Forest Service ranger, found the Hayman Fire while patrolling during a fire ban. Later, she admitted to starting it, claiming she was burning a letter from her husband denying her a divorce. Investigators, however, found no evidence of burnt paper in the ring where the blaze started, Barton's husband denied writing the letter, and she had recently asked to attend a wildfire investigation class, leading to suspicions that she started the blaze to make herself into a hero by discovering it.[5]

During Colorado's explosive fires of 2012, firebugs again plagued the area. The Teller County sheriff reported that a half-acre blaze outside Woodland Park was suspicious, while during a single two-hour period that same day firefighters in Divide responded to seven arsons in the forest. In Colorado Springs, a bastion of conservative politics, the arsons in the forests near the city sparked chatter of terrorism in coffee shops and online. FBI agents joined with Colorado Springs police to investigate them.

"It infuriates me and it just makes my blood boil," Colorado governor John Hickenlooper said of the possibility that the Waldo Canyon Fire was arson. "It creates a physical reaction in me."[6]

Wildland arson was increasingly seen as a serious crime. Three years earlier, in California, a jury sentenced arsonist Raymond Lee Oyler to death after convicting him of murder for starting the Esperanza Fire that killed five firefighters on a U.S. Forest Service engine crew in 2006. Oyler was the first person sent to death row in the United States for starting a wildfire, but just three years after his sentence, prosecutors asked for the death penalty for Rickie Lee Fowler, who was convicted of starting a 2003 wildfire in the San Bernardino Mountains blamed for five fatal heart attacks.

ARSONS, IN REALITY, make up a small portion of the wildfires humans start. Research from Jeffrey Prestemon of the U.S. Forest Service and economist David Butry showed that arsons in national forests declined by more than 80 percent per capita between the late 1970s and 2008. Fires ignited by discarded cigarettes, another popular scapegoat for wildfires, declined by 90 percent. Railways, which once started myriad wildfires with sparks from their wheels and engines, also start fewer fires today, largely due to spark arrester technology.[7]

Still, as communities and roads encroach deeper into wildlands, human ignitions, most of them unintentional, increasingly outstrip natural starts. Power lines, trash burns, campfires, and sparks from engines, wheels, and machinery ignite thousands of blazes every year. In February 2017 Jennifer Balch, a fire ecologist at the University of Colorado, and several of her colleagues released a paper that looked back over 20 years of government agency wildfire records in the United States. They found that between 1992 and 2012, humans started 84 percent of the nation's wildfires. Humans also tripled the length of the fire season by starting those fires at times of the year when lightning ignitions were rare, and spread those blazes over an area seven times larger than that affected by natural blazes.[8]

One investigator I met described a suspicious fire that ignited along a roadway. He dug through the charred weeds, certain he'd find some type of incendiary device, but instead discovered a culprit that left him shaking his head and laughing. A broken bottle had magnified a sunbeam to ignite flames in the parched grass.

PART V

EXTENDED ATTACK

20

MOUNTAIN SHADOWS

Colorado Springs — June 23, 2012

FROM HIS TRUCK RACING SOUTH on Interstate 25 on the Saturday in which the most wildfires on record exploded across the state, Jim Schanel could look down the length of the Front Range of the Colorado Rockies. The plume of the High Park Fire above Fort Collins filled his rearview mirror. His windshield framed the billowing cloud of the Waldo Canyon Fire outside Colorado Springs.

"This is beyond what mankind can deal with," he thought. "This is going to be catastrophic. We don't have the resources [or] any idea how to manage it."

For more than 30 years Jim drove at columns of smoke with the adrenaline rush of a kid on Christmas morning. "This is going to be a good firefight," he would think as he approached big ones. This time, however, he had a knot in his stomach.

"I saw the header in my rearview from High Park and a header in my front windshield," he said. "[For] the first time, I think in my career, I was terrified."

During the previous 13 days at the High Park Fire, he'd fought to save homes scattered throughout the mountains. On his last day on the blaze above Fort Collins, flames climbed out of Poudre Canyon and charged north at Red Feather Lakes and the Glacier View Meadows subdivision. "We were hoping it wouldn't," Schanel told me. "But it did."

He was working with a federal team on structure protection, side by side with state, county, and local crews, as well as National Guard troops. "We put a lot of resources in there," he said. But the fire behavior was like nothing Schanel had ever seen. "We pulled a lot of firefighters back to safety zones," he continued. "We lost some structures, and we went back in after the fact and we were able to save some."

When Schanel could get a view to the south, he could see the smoke rising above his hometown.

Schanel, a structure protection specialist on the federal Type 1 Incident Management Team fighting the High Park Fire, was also a battalion chief on the Colorado Springs Fire Department. Burly, with a graying mustache and worn-out knees, he was the city's most experienced forest firefighter. He still had a day left in his two-week assignment with the feds when he told the High Park leadership that he needed to go home. "My backyard is having their day," he told them. "It's . . . the big one."

But it wasn't the only big one. Multiple large fires were going to make getting resources for the new ones difficult. There already weren't enough planes or retardant to drop from them.

"If there's no slurry, what do you do? You can't put firefighters in front of that kind of extreme fire behavior," he said.

"Driving down the Front Range and seeing that header, I knew we were in trouble and that we didn't have the infrastructure in place to handle it," he continued. ". . . A lot of fires were going to go unattended or unchecked. I knew the behavior of these fires and the condition of the fuels. They were going to become big and fast and impact a lot of people."

THE WESTERN EDGE OF COLORADO SPRINGS is about 10 miles due east and 8,000 feet below the summit of Pikes Peak, the most eastward of the seventy-six 14,000-foot mountains in the United States. The peak is more than twice the height of any mountain between it and the Atlantic Ocean. At its foot the forests are decidedly drier than those at the north end of the Front Range, with oily and fire-prone Gambel oaks mixed among the ponderosa pines. A series of ridgelines and wooded canyons descend into the city of nearly 440,000 people. Colorado Springs presses so hard against the mountains that one-third of it is in the wildland-urban interface, where homes are threatened by wildfire.

Waldo Canyon cuts into the mountains about a mile northwest of Manitou Springs, a quaint hamlet of artists and tourists that is effectively the western arm of Colorado Springs. Another gorge, Williams Canyon, rises due north out of Manitou like a long funnel. Three miles to its northeast is Queens Canyon, which spills southeast from the mountain slopes to Glen Eyrie, a Christian conference center and castle, and then onto the Garden of the Gods, the red, 300-foot-tall fins of rock that mark the western edge of Colorado Springs like the relics of an ancient wall.

Mountain Shadows, a suburban neighborhood dense with housing on wide, paved streets, lies within that imaginary wall just north of Queens Canyon. Even though the subdivision abuts the Pike National Forest, most of its residents didn't consider wildfires a threat. Most of the neighborhood's trees had been planted there by its residents, and lawns provided most of the grass. It seemed impossible that a fire could burn downhill from the mountains into the suburban development. To add to their sense of security, just north of the suburb is the 18,500-acre grounds of the U.S. Air Force Academy.

Cedar Heights is outside the imaginary wall — a half mile west of the Garden of the Gods, and the same distance east of Waldo Canyon. Dozens of roads, at least 15 of them dead ends, burrow through the woods atop a foothill between Williams and Queens Canyons. As opposed to Mountain Shadows, Cedar Heights seemed an easy target for the fire.

WITHIN HOURS OF TRYING to hold the High Park Fire back from the homes of Glacier View Meadows, Jim Schanel was 170 miles south preparing Cedar Heights for the arrival of the Waldo Canyon Fire. Bulldozers carved a line between the development and the forest. Planes painted the edge of the evacuated neighborhood with retardant. Teams set up trucks and hoses to protect homes, a microwave tower, and infrastructure for the city's water department.

I joined spectators crowding near the Garden of the Gods to watch bombers dropping slurry as 100-foot flames leapt close enough that we were convinced the clouds of black smoke came from burning houses. But late Monday the fire retreated without having burned any homes. The work put in by firefighters and residents preparing their properties and neighborhood — Cedar Heights was one of the most "firewise" communities in Colorado Springs — paid off. The fire burned northwest, away from the city.

"People kind of relaxed a little bit," Jim recalled. "It wasn't right at our doorstep anymore."

At a community meeting, evacuees demanded to know when they'd be allowed back into their homes. Other residents called the county to complain about neighbors who had allowed branches, pine needles, and weeds to accumulate in violation of fire codes. Monday's temperature reached 98 degrees, the fourth record-hot day in a row.

"If you looked at what was going on in the rest of the Front Range, we probably should have been hypervigilant the entire time that fire burned away from us," Schanel told me.

RICH HARVEY IS ONE of 17 Type 1 incident commanders in the nation. His day job is with the Nevada Division of Forestry, but he puts that aside when the federal government taps him to manage the response to the largest, most complex fires in the United States.

The National Incident Management System categorizes fires from Type 5, which can be taken care of by just a handful of people and a truck or two, to Type 1, which usually has more than 1,000 personnel as well as aircraft and support equipment. Management runs from the National Interagency Coordination Center, the huge war room in Boise, Idaho, watching over the responses to fires across the country, through 11 regional coordination centers, and down to state and local governments. Harvey was called up by the feds in Boise, but he would have to work with the locals who had already responded.

Lanky and folksy, he started fighting fires 35 years earlier, when Colorado was known for its "asbestos forests," which rarely burned big enough to require a visit from a Type 1 incident commander. Now the state was having so many big fires, it seemed like Harvey's second home. In 2012 he fought fires in Colorado from March until December, starting with the Lower North Fork Fire, the controlled burn that ran wild and killed Ann Appel and Sam and Moaneti Lucas.

"The Lower North Fork was surprising," he told me. "It's March and you're at 8,000 feet . . . That's not normal. It was just too early and too high of a place."

But the Waldo Canyon Fire, three months later, was in the middle of Colorado's fire season.

Harvey and his crew of about 50 wildfire managers took over the blaze at six Monday morning. He could see that all the windows of opportunity for a bad fire were wide open and lined up, just waiting for a strong wind to blow through them. Record temperatures and extreme drought had stressed overgrown forests of ponderosa pines. Oily and fire-prone Gambel oaks had been nipped by a late frost, killing their flat leaves, which now rode the wind like flaming papers to start spot fires. High-elevation forests of lodgepole pines had historically low moisture content and tended to burn in "stand-replacing" conflagrations.

"Colorado had a bull's-eye on it for fire potential," Harvey said.

There was another complication that he had never seen in Colorado before. The fire was running at a city of nearly half a million people.

"Every direction the fire [could] move in [was] a bad direction," he told me.

Harvey started trying to connect the dots that could keep the fire out of Colorado Springs. Bulldozers and hand crews dug a fire line between the forest and the city, anchoring it in the already burnt land of Cedar Heights. He sent other crews and engines up Rampart Range Road, which switchbacks up a ridgeline northwest of the city into the mountains. Air tankers could help hold the ridge with retardant drops. But if the fire made it past the road, it had an open run at Colorado Springs. And once it hit the city, there didn't appear to be much he could do about it.

Harvey's command ended at the city line. Colorado Springs is a "home rule" municipality. If the fire crossed into the city, Mayor Steve Bach and his fire chief, Rich Brown, announced, the decisions to evacuate and how to fight the blaze were theirs.

Steve Riker, a captain with the Colorado Springs Fire Department, ended up as the incident commander inside the city. Riker started his firefighting career in the U.S. Air Force two years after Harvey started fighting wildfires. His father had been a Marine Corps firefighter at

Pearl Harbor. Among the firefighting memorabilia on the mantel of Riker's home is a black-and-white photo of his father carrying a young woman out of a burning house in Flint, Michigan. That woman died, but he recalls another woman whom his father saved when she was a child calling him more than 20 years later to thank him. "That's the only time I've seen my dad cry," Riker said.

The captain was a heavy rescue specialist, but had no current wildland firefighting certification. And despite his experience and the long lineage of firefighting in his family, Riker had advanced as far as he could in the Colorado Springs Fire Department because he didn't have a bachelor's degree. Still, he would hold the position of incident commander on the largest fire in Colorado Springs history.

Riker was on his way to officiate a basketball game when he heard that a fire had broken out in Waldo Canyon. He left the gym to oversee a handful of engines that patrolled the western edge of the city looking for firebrands or embers that could start spot fires. He could see flames from the growing wildfire in a canyon. It was as close to the city as he'd ever seen a forest fire. That night, while they watched over the Glen Eyrie estate, his crew bedded down wherever they could to grab a bit of rest for the next day.

On Monday, while he was patrolling, residents in Mountain Shadows asked him if he thought they should evacuate. "I would have left 24 hours ago," he told them.

ON MONDAY AND TUESDAY MORNINGS Riker and his firefighters held their briefings jointly with Harvey's Type 1 team. After the briefings, however, the two teams split up and worked independently. The federal firefighters confronted the blaze on the mountain, while the city firefighters set up to keep it there. This is just one indication of the increasingly complex and confused response.

The city had no experience with this kind of fire. While most of the firefighters on the west side of Colorado Springs were trained to fight forest fires, only about half the city's firefighters overall had "red cards" certifying that they had trained for wildland firefighting. Communications, maps, staging areas, and mobile command posts would all have to be arranged on the fly.

The division between the federal and city firefighters was just the beginning of the jurisdictional complications. El Paso County sheriff Terry Maketa was the county fire marshal, so he managed evacuations and the firefighting response there. Air force firefighters would protect the academy. Even the Colorado Springs Utilities had its own fire crews and command structure.

The different jurisdictions had different types of radios that did not necessarily communicate with one another. When Riker picked up a vehicle at the start of the fire, he was initially given a maintenance truck with no radio at all. He switched that vehicle for a Tahoe that had a radio and a lot more power. But it didn't have a map book.

Maps would be a challenge, whether there was one in the truck or not. "We have 21 fire stations," Riker said. "I doubt you'll find two map books that are identical [between them]."

He stopped at a fire station with a large printer and made several copies of the station's large map of the western edge of Colorado Springs. At least he and his subordinates would be on the same page.

Diana Allen, a fire behavior analyst, arrived as part of the Type 1 Incident Management Team on Monday and immediately saw that the fire was launching embers up to half a mile away. Instead of cooling off that night, she said, the temperature remained high and the humidity stayed low. Firefighters working overnight reported that the blaze raged right through the dark, heating and drying the fuels that it didn't consume. They would ignite easily the next day. The Haines Index, which measures how moisture and stability in the lower atmosphere drive fire behavior, was at a 6 — the highest level. Early Tuesday the National Weather Service issued a Red Flag Warning, and at that morning's briefing Allen warned firefighters to expect extreme fire behavior that afternoon.[1]

Eyes across the country were on Colorado Springs. In California, David Blankenship, a former employee of the city's fire department, was watching the fire through a software system he designed that analyzes satellite images to predict wildfire behavior for the U.S. Forest Service. For the first time in the years that he had used the program, the very pixels seemed to be shouting a warning. "Something bad is going to happen," he said.[2]

But not everyone was getting the message. The city had yet to call in additional firefighters from other municipalities, and had no maps to give them if it did call them in. City leaders had just that morning started drafting an evacuation plan for the northwest corner of the city. And while firefighters were briefed about the potential for the fire to burn into Colorado Springs, city officials announced that residents who had been evacuated over the weekend, including those in the Mountain Shadows neighborhood at the edge of the national forest, could make visits back to their homes.

THEY WEREN'T THE ONLY ONES HEADING HOME.

Jim Schanel had only taken catnaps since arriving in Colorado Springs that Saturday, after two weeks spent fighting the High Park Fire, so Tuesday morning commanders sent him home.

But he couldn't sleep. From his house he had a panoramic view of the city and the mountains. A pyrocumulus cloud, formed when the superheated column of air blasted burnt particles high enough into the atmosphere for water to condense on them, piled up thousands of feet above the mountains. The thunderhead teetered on top of the smoke column, leaning toward the city.

Fire clouds can bring rain, which may assist firefighters but usually falls far from the blaze, or they can start new fires with lightning strikes. Their greatest threat, however, is when the column supporting them collapses, and the superheated air that was aimed skyward blasts horizontally across the landscape. Schanel hadn't seen many column collapses early in his career, but had noticed more of them recently. Increasingly unstable atmospheric conditions, combined with heavy fuel loads that sent more particles into the air, created tippy, top-heavy smoke columns.

"That smoke . . . [is] so inundated with millions of tons of carbon particulate and needles and shingles and yard debris and all those things that the convective column pulled up," he said. "It [doesn't] have enough thermal energy to keep itself aloft."

Ten miles to the west there were other clouds. A storm cell was building over the 10-year-old burn scar from the vast Hayman Fire, and winds flowing out from it pressed against the smoke column above Colorado Springs. "The indices were setting up for that fire to take a big run," Schanel told me.

He kept waking up to look at the fire. Around 2 p.m. he could see the weather starting to spin behind it, and by 4 p.m. he could see the fire making a hard run at the city. He put on his fire gear and headed back out.

THAT MORNING, TUESDAY, I stopped off at the High Park Fire's press briefing, then drove to Rocky Mountain National Park to tour the neighborhood that had burned there the weekend before. As I walked through the cinders of the razed houses, my phone rang with a text from my wife. Yet another new fire had broken out, this one on Flagstaff Mountain above Boulder. Pre-evacuation notices had gone out to the neighborhood where we lived. "Just find the cat," I texted her, and started driving back south.

Another Type 1 Incident Management Team had just arrived in Colorado and barely had their feet on the ground when they were assigned to Boulder, along with an air tanker that firefighters in Colorado Springs had hoped would be working the Waldo Canyon Fire. Suppression efforts are most likely to succeed immediately after a blaze starts and is still small, so dispatchers prioritized getting resources onto the fresh fire. By the time I returned to Boulder, air tankers were bombing the fire behind the Flatirons. I organized the most critical of our files and photos in case we needed to move out of our house quickly, then photographed the blaze above Boulder. From the National Center for Atmospheric Research, below Bear Peak, I could see the flames on the mountain.

Within a few hours the Flagstaff Fire was under control, and I was back on the highway, heading to Colorado Springs — the fourth wildfire I'd visit that day.

21

FIRESTORM

Colorado Springs — June 26, 2012

DURING THE EARLY AFTERNOON ON TUESDAY, the Waldo Canyon Fire steamrolled toward Rich Harvey and the firefighters holding the blaze west of Rampart Range Road outside the city. Spot fires popped up a mile or more ahead of it. Firefighters reported that a flaming deer jumping across the road ignited one of them.

At the massive High Park Fire still burning above Fort Collins, the Horsetooth Reservoir kept the fire from the city. But even a lake wasn't stopping this blaze. "It spotted across the Rampart Reservoir," Harvey told me. "It spotted across the reservoir!"

As the commander in the mountains tried to slow the fire's push to the north, Steve Riker, the commander in the city, moved some of his resources in that direction. His crews and trucks ran parallel to the fire's progress like defensive backs mirroring a shifting offense before a football is snapped. The city limits were effectively the line of scrimmage. Riker and his crew would engage the fire as soon as it crossed the line.

But there weren't many resources to move. He had only four fire trucks to protect everything from Mountain Shadows to the Air Force Academy — subdivisions with thousands of homes. At the edge of Mountain Shadows, Riker noted that Chipeta Elementary School would be the best safety zone for firefighters to fall back to if things went to hell.

Then he looked up warily at a quarry above Queens Canyon. "Where we have that rock pit," he said, "that's where we'd probably see it come."

The two incident commanders — Harvey with the feds and Riker with the city — were like generals from neighboring nations fighting a common enemy on different fronts. They weren't in contact, but they had the same worry. If the fire made it past Rampart Range Road and into Queens Canyon, it could quickly climb onto the ridge above Mountain Shadows.

"Queens Canyon was a trigger point [for evacuations]," Harvey said. "If the fire becomes established in there, there is no good containment, no good place for making a stand between Rampart Range Road and Colorado Springs . . . There are no roads, no trails, no natural barriers."

By noon the city had shut down visits to previously evacuated neighborhoods, but few residents who had gone home were aware that they needed to leave again. Just before 2 p.m. the city issued pre-evacuation notices for Mountain Shadows and the Peregrine development north of Chuckwagon Road.

A few minutes after the pre-evacuation notice was issued, a lookout posted near the Queens Canyon Quarry scar reported that the fire was approaching the canyon. "It's the furthest east and north we've seen activity so far, and it is very active," she reported.[1]

At 2:17 p.m. another lookout standing atop the quarry radioed that the fire was "starting to drop down into the canyon."

At 2:40 he saw "burning material rolling down into the canyon."

"The fire is getting down into the bottom of the canyon," he reported. "Really heavy fire activity on the western lip of the canyon."

The fire was moving down the canyon, the lookouts reported, and was climbing up the other side.

"We have flames on the ridge," a firefighter called at 4:06.

Seconds later Steve Riker reported "heavy fire" above Mountain Shadows. "It is just running hard right now."

A report that the city was in danger went out to everyone: "All units, I have fire on the ridge . . . We need to put that plan in place."

Despite the fire passing the trigger point for evacuations, the call to start them didn't go out.

THROUGHOUT THE DAY Riker had grown increasingly concerned that Mountain Shadows wasn't yet evacuated. At a house below the ridgeline, Riker asked the homeowner if he could use his back deck to watch the fire. The man threw him his keys. "Lock it up when you leave," he said. "We're out of here."

"Flying brands were moving about a hundred yards from the main body of fire," Riker said. "And then that area would fill in [with flames] in a matter of minutes."

Riker called in a task force — six engines and four brush trucks — that he had waiting to meet the fire where it entered the city. Despite its rapid spread, Riker thought he could stand up to it. The winds had been low, maybe 5 miles per hour, and there was a flat, grassy meadow between the forest and the homes.

"I'm going to fight this fire," he thought. "I've got 200 feet of level ground and the wind's not pushing, and I'm going to be able to do this."

He stationed the engines and brush trucks along the streets that would first encounter the fire and had them deploy their wildland hoses, which are lighter and easier to carry and pack up than structure fire-fighting hoses. Wildland firefighters usually connect their hoses only to trucks or portable pumps, in case they need to get away fast, but Riker encouraged anyone who wanted to tie into a hydrant to do it. One thing Colorado Springs has, due to its position at the base of the Rockies, is terrific water pressure. He told his firefighters to spray as much as they needed to.

"Push comes to shove," he said, "if you need to leave that hydrant, that's what we have an axe for. You're just going to hit that hose and leave."

Mike Myers, the chief of the Colorado Springs Utilities fire crew, had been working side by side with Riker through the day, and Lieutenants Bill Pellegrino and Steve Wilch of the Colorado Springs Fire Department joined them to prepare for the fire's arrival.

"It's about a third of the way down, and it's starting to move a little bit fast," Riker relayed to his men over the radio.

A FEW MINUTES BEFORE THE FLAMES HIT the ridge above Mountain Shadows, Mayor Steve Bach was in an interview with CNN in which he reported that the city was not in danger and businesses were open. The city's biggest sporting event, the Pikes Peak International Hill Climb, in which race cars from around the world fishtail up switchbacks to the mountain's summit, would go on as planned two weeks later, the city announced, as would a two-day festival in downtown Colorado Springs before the race. In the end, the race would be delayed by more than a month.

Ten minutes after the mayor's interview, at a press briefing in the parking lot at Coronado High School, Bach joined a lineup of federal, state, county, and city officials that reflected the increasingly complicated nature of the response.[2] When they weren't at the microphone, they looked over their shoulders at the flames on the mountainside. Outflow winds from the storm cell building behind the fire tipped the smoke column toward the city.

Jerri Marr, the U.S. Forest Service supervisor for the Pike National Forest, responded to a reporter's question: "I'd say that we're still on the offensive today."

Incident commander Rich Harvey stepped up to the microphone. "In some of those places, we're still on the offensive," Harvey said. "In others . . . we're transitioning from offense to defense."

AT 4:13 RIKER RADIOED from Mountain Shadows, where the flames were closing in. "We did do mandatory evacuations?" he asked fire battalion chief Ted Collas.[3]

"Negative," Collas radioed back. "I will confirm that and make sure it is happening."

The fire, Riker noted, was about halfway down the slope above Mountain Shadows and moving fast. "It's coming down the side of that hill," Riker called back. "I've got spot fires moving in below."

AT THE PRESS CONFERENCE aides to Mayor Bach beckoned him into a huddle away from the other officials while Rich Harvey took questions.[4]

"What are the consequences of the fire getting into Queens Canyon and if it keeps running?" a reporter asked.

"That's a good question and a tough answer," Harvey responded. "Rampart . . . Range Road was one of the best . . . high-probability success points for the containment of this fire."

Mayor Bach stepped back to the microphone and cut off the commander. "We've just been told that there is now an evacuation order for the balance of Mountain Shadows up through Peregrine," he said. "Please get that on your newscasts right away."

"Is that mandatory or voluntary?" a reporter shouted.

"Mandatory!"

The first evacuation calls to upper Mountain Shadows came at 4:24, nearly 20 minutes after the fire hit the trigger point. A second round went out at 4:37.

AT 5:11 THE STORM CELL 10 MILES WEST of Colorado Springs splashed gusts of wind over the Front Range.[5] They pushed hard on the smoke column teetering toward the city. As it leaned over ground that wasn't burning, the smoke no longer got lift from the heat of the flames. The column collapsed onto Mountain Shadows like a chimney falling down. The superheated air that had been rising from the fire into the sky blasted down the slopes onto the city. Thousands of airborne embers, some weighing more than a pound, showered the neighborhood. Winds gusted more than 65 miles per hour, and the fire came down from the mountains like an orange avalanche.

DURING THE HAYMAN FIRE 10 years earlier, Cindy Maluschka and her husband, Mark, lived in Woodland Park, in a forested bowl on the other side of the Rampart Range. Although that fire only came within about six miles of their home, it was terrifying enough for them to rent a storage unit and fill it with the things they couldn't bear to lose. "We kind of did a little move," she told me. "We took boxes and boxes."

But when they moved to Mountain Shadows, they had little fear of wildfire. "There were more streets and more services, and the distance from the forest itself," she said. "We certainly felt safer from a wildland fire."

Most of Cindy's friends and neighbors didn't even know the name of Queens Canyon. The initial evacuation zone ended at Chuckwagon Road, two streets away from them. "Once that boundary line was drawn," Cindy said, "it created . . . a very false sense of security."

Still, she and her husband gathered documents, family photos, and the tooth their daughter, Amber, had recently lost. They put a clothes-basket in the little girl's room. "Pack this with all your most important things," her mother told her, limiting the number of stuffed animals the six-year-old could bring with her.

They took a fraction of what they had stored during the Hayman Fire and moved into Cindy's father's house, far from the evacuation zone. Three days later, when city officials announced that residents who had been evacuated could visit their houses, Cindy and Mark returned home to get some work done on their computers. Their house was still outside the evacuation zone.

"There was absolutely no more urgency on that day," she said. "We weren't on mandatory evacuation. We weren't even on pre-evacuation."

At one that afternoon much of Mountain Shadows was again given a pre-evacuation notice, but few of the residents who had returned to their homes got the word. When the clouds turned orange during the afternoon, Cindy thought it was just the sun reflecting off them. She watched the 4 p.m. news conference on television and saw nothing to worry about until the mayor's sudden announcement that all of Mountain Shadows was on mandatory evacuation. Reverse 911 calls went out, but many residents who had signed up for the service didn't receive one.

"I ran up the stairs and opened the garage door, and the mountain was on fire," she said. "I felt the heat. I felt . . . kind of like tornado pressure."

They had 10 minutes to pack the last things they would save — computers and clothes — and get out. Outside the house Mark shouted to his wife that he needed to go back in to rescue the family's goldfish.

"I am not going to do this for goldfish," she replied as she drove away, leaving him to make his way out with the couple's other car. "Screw the goldfish."

Racing off, she was stunned to see many of her neighbors standing in their driveways, gazing at the flames. "You have got to leave!" she screamed at one of them. "Quit taking pictures!"

When her husband caught up with her at her father's house, he had the goldfish and a photo of a wall of flame coming down their street. "I knew, absolutely, that our house was going to be gone," she said. "I didn't see how it could not be."

The Colorado Springs Police and Fire Departments evacuated some 26,000 people on Tuesday, most of them after the 4 p.m. press conference.[6] By nightfall evacuations would extend from Cedar Heights all the way to the housing at the Air Force Academy and include 32,000 people.

In the minutes after the mayor's announcement, front doors and garages opened simultaneously on densely developed suburban streets throughout Mountain Shadows. Flames leapt on the ridgeline above the rooftops and splashed down the hillsides. Residents ran to their cars with boxes, children, and pets. Others dashed to the homes of neighbors and pounded on the doors. A few pulled away in their cars with their homes already burning.

As the smoke column collapsed over the city, it turned the bright afternoon dark. Chuckwagon Road and Centennial Boulevard were backed up with hundreds of cars, their lights on to cut through haze the color of orange juice. Initially only pine needles and tree bark fell with the ash and embers, but soon roofing shingles, wood siding, and paper rained down.

Michael Duncan had come home from work earlier that afternoon, after a friend called him to say the fire was threatening the city. He made a phone call for his job around 5 p.m., but when he finished, he and his wife and children could see fire up the street that leads from the mountains straight to their house. The rest of his family evacuated, but he stayed, hoping he could save their home. He hosed down the roof. His daughter, Peri, called him repeatedly asking him to retrieve photo albums, then her brother's guitar.

When the fire was half a lot away, the smoke got so dense he couldn't see up the street. The smoke alarms throughout the house went off. "I decided I didn't want to stay and die to that kind of noise," he told me.

As he pulled away in his truck, all he could see along the street were pine trees torching and homes burning.

THE COLLAPSING COLUMN LANDED on Steve Riker and his firefighters.

"It went from 5-mile-per-hour winds to 60 ... in a matter of minutes," he said. "And that [turned] a two- to three-foot brush fire that I was looking at into a massive fire coming in our direction."

A helicopter made a water drop about 200 feet in front of them, then banked over them. The pilot made eye contact and swirled a hand in the air.

"I think he wants us to get out of here," Riker told the chief of the Colorado Springs Utilities fire crew, who was standing beside him.

"Yeah, that's exactly what he's telling us to do," the chief responded.

They ran back from the flames to the relative safety of the pavement.

"I had firebrands the size of a fist fly past my head," Riker said.

He called for all of the firefighters to retreat about a quarter mile to the Chipeta Elementary School safety zone he'd scoped out that morning. None of them had to chop their hoses loose from a hydrant with an axe, but a few left their hoses behind altogether. They headed up streets clogged with evacuating residents. Riker was the last to leave. Just before he climbed into his rig, he looked up and saw 100-foot-tall flames "as far as I could see north and south."

His visibility dropped to 100 feet and then to 50 feet. When he finally started driving, he couldn't see past the hood of his truck. He tried not to rear-end the fire truck in front of him.

The fire came into the city in three phases. First, the fire front ripped through the trees, scrub, and grass. Then an ember attack showered firebrands miles ahead of the front. Finally, the fire leapt from house to house. Entire blocks went up, with each burning house spreading flames to the next one down the street. Some homes ignited deep in neighborhoods where nothing else was burning.

"We're talking about an urban conflagration," Riker said. "Going from house to house . . . going from back deck to back deck."

He was terrified that a firefighter had become trapped between the fire front and the dozens of spot fires and burning homes, become lost in the darkness of the smoke, or gotten stuck in a traffic jam that had been overrun by the fire. During their retreat police cars passed them, racing back into the neighborhood. Out-of-town relatives called the police about family members they thought were still in their homes, and officers were knocking on doors to check.

"There's nothing we can do at this point," Riker called over his radio to the police. "If you're . . . in that area, you're going to die."

When the caravan of firefighters reached Chipeta, Riker was ecstatic that he could account for everyone on his rosters, but his relief didn't last long. When he turned around from his roll call, the heat from the firestorm slapped his face.

"It's right on us again," he said. "We've got to leave this area!"

The entire force retreated again, this time to an old telecommunications building at the intersection of 30th Street and Garden of the Gods Road. But again the fire was right behind them. They'd only been there a few minutes when 30-foot flames rose in the thick vegetation behind the building, one of the dozens of spot fires sprouting throughout the neighborhood. The firefighters' third evacuation would have been to Station 9, more than two miles into the city. Riker began to worry that the fire would threaten his own home, some eight miles from Mountain Shadows. "I didn't think we were going to stop this thing," he said.

He turned around to see six green buggies of hotshots from the U.S. Forest Service behind him. He didn't know where they had come from. Rich Harvey had sent crews from the firefight he was leading in the mountains to assist the firefighters in the city.

"Can you guys stop this right here for me?" Riker asked the hotshots' supervisor, referring to the spot fire that was threatening them. "Can you put a line down and stop this?"

The hotshots had no water, but that didn't seem to worry them. They ran at the blaze with Pulaskis and chainsaws, digging and dropping trees like badgers and beavers. In minutes most of the field was just a patch of dirt. The spot fire died.

"The most amazing thing I've ever seen . . . in my career," Riker said.

They wouldn't be retreating again.

Riker took the large map he'd had printed earlier and taped it to the hood of his truck. Then he started writing lists and notes on the truck. In the next few hours the notations would run like hieroglyphics from the map, across the hood, and down the fenders of the Tahoe.

FOREST FIREFIGHTERS SPILLED IN from the smoking mountain, and municipal fire crews arrived from cities and towns all along the Front Range. The manager of the Loaf 'N Jug, a convenience store on Garden of the Gods Road, opened up for a Forest Service crew from the Tahoe National Forest that had fought fires from Alaska to the Australian Outback.[7]

"Where are your maps?" a firefighter asked, and then took every one of the area that the store had.

Other Forest Service crews were lost on Mountain Shadows' byzantine roads, often working alone with no time to pick up their radios.

Jim Schanel had trouble getting back into Mountain Shadows — traffic evacuating the neighborhood clogged both sides of the streets. When he finally managed to weave through the oncoming vehicles, he met up with Steve Riker and his firefighters as they regrouped. Riker had tears in his eyes when Schanel saw him.

"I think we lost people," Riker said.

Schanel was sure he was right. Hundreds of homes were burning. Many of them looked like they were still occupied, with two cars in the driveway or lights on.

Riker, Schanel, and Mike Myers, the Colorado Springs Utilities fire chief, gathered around the map taped to the incident commander's truck and divided up the western edge of the city into three sections. Each of them took a piece. They couldn't stand up to the fire front any more than they could stand up to a tornado or a tsunami. So they would follow behind it, saving what they could after the leading edge of the conflagration passed.

The first crews Riker sent to try to keep the flames from passing 30th Street saw they would be overrun there and pulled back.

Schanel headed into the most northerly section of Mountain Shadows and the Peregrine neighborhood above it. He tried to steel himself as he looked up each burning street he passed. "You're going to see an engine company burnt over," he told himself.

Schanel led about 18 of Colorado Springs' largest fire engines, several smaller engines, and a couple of strike teams from Denver, as well as the Redding, Vandenberg, and Tahoe Hotshots and some other Forest Service crews.

Colorado Springs fire trucks carry both heavy structure firefighting gear and the lightweight Nomex firefighting clothes of wildland firefighters, but few of the crews that arrived from other cities and towns were trained or equipped to fight both a structure fire and a wildfire. None of them had ever fought a fire like the one they were facing. In addition, the Colorado Springs firefighters had no maps to give the visiting crews.

The structure firefighters' oxygen tanks would run out within an hour, and their heavy gear made it impossible for them to quickly chase down or escape a fire spreading through the neighborhoods' landscaping. Wildland firefighters were fast and nimble and could go for eight hours straight, but they had nothing to help them breathe in the heavy smoke of a structure fire. They'd work until their vehicles were running out of fuel or water and discover they had no place to re-fill. The different crews' radios couldn't talk to one another. Even their hose fittings were different, so they couldn't tap into one another's water trucks or pumps.

"Communications, command and control, interagency doctrine, standard operating procedures, I mean, pick something," Schanel said. "[We have] completely different agencies with different philosophies."

But if they didn't stop the fire in Mountain Shadows, it would spread through the city.

"We need to do this or die trying," Schanel told his crews. "We're the only people here right now. This is all we have, and if we don't stop this, it's gonna be beyond devastating. Let's go to work."

Schanel built his tactics around each team of firefighters' training and equipment. He positioned his largest fire trucks from the city at hydrants. They sprayed down burning homes to try to reduce the heat that was pushing firefighters back and igniting other houses. The wildland crews chased down spot fires that were igniting a mile or more from the burning homes.

A crew from Denver had no wildland training or gear. Their battalion chief still had his tie and badge on. "I thought we were going to back up a few of your stations," he told Schanel. "I didn't come for this."

"Well, Chief, we've got hydrants," Schanel replied.

Don't think of it as a wildfire, he told him. "It's like we had a gas explosion," he said. "You could take that block."

When he returned an hour later, the Denver crew had knocked down the fire there and kept it from spreading.

Riker split up his Colorado Springs firefighters with wildland training and knowledge of the city to assist the out-of-town crews. One Denver crew was about to be burned over when the Colorado Springs firefighter with them had them retreat.

Lieutenant Dave Vitwar, who, like Jim Schanel, had been working on the High Park Fire until the Waldo Canyon Fire threatened his home turf, had one of his Colorado Springs firefighters assigned to each truck of a strike team that arrived from Denver. They found streets where every home had burned and collapsed in on itself, leaving only a foundation filled with rubble. They abandoned those streets, but if a street had only one home burning, the strike team surrounded it and sprayed curtains of water to keep it from igniting other houses. On one street where a single home survived, Vitwar noticed a puff of smoke come out of the bricks like a snake's tongue, then get sucked back in.

"Chief, you see this?" he asked the strike team leader working with him.

Firefighters cut through the garage door and tore into the drywall inside to extinguish the hidden flames.

At the edge of Mountain Shadows closest to the forest, the fire front had burned every home. But elsewhere in the neighborhood, the destruction was random. Blocks burned to the ground inside neighborhoods where no other homes ignited, as if bombs had gone off in the middle of the subdivisions.

The Denver firefighters, still working in their heavy bunker gear, were getting spent after an hour or two on the fire. Vitwar had them all pull the insulation out of their fire coats and pants, leaving only the flame-retardant shells. They were going to be at this for a long time.

Other crews found themselves on roads with fires burning below them, the most dangerous place to be in a wildfire.[8] Escape routes and safety zones on the winding streets were often hidden by the dense smoke. Some decks were surrounded by scrub oak, and houses were so close together that firefighters in between them couldn't extend their arms. Flaming houses teetered like they were made of cards, threatening to fall on the ones beside them. Wooden siding ignited just from the radiant heat.

The federal wildfire crews' safety protocols forbid them to enter burning buildings, and they didn't have the oxygen equipment or heavy bunker gear needed to fight structure fires. But they could pull fires away from homes before the flames ignited them. Hotshots with chainsaws cut down fences carrying the fire and sawed burning decks off houses. One of their supervisors pulled his truck into a backyard, wrapped the cable from its winch around a burning deck, and ripped it free of the house.

Firefighters grabbed garden hoses and ladders they found outside houses and put them to use fighting the blaze. When they tripped over sprinkler systems or hoses in yards, they turned them on and left them running. Others pulled propane tanks from barbecue grills and threw them into the street, away from the flames. Exploding tanks at the homes that were already burning blasted into the sky and landed with a thud.

STEVE WILCH, A LIEUTENANT with the Colorado Springs Fire Department, had seen wildfires burn into cities in Southern California, where he grew up and learned to fight fires. Although that had never happened in Colorado, he knew it was just a matter of time.

On Tuesday he was working with a team that bulldozed miles of fire lines around Mountain Shadows but had more to go when they saw flames leaping on the ridgeline above them. They were out of time.

"Game on," Wilch thought.

Within an hour he and the dozer crews joined Riker's retreat to their safety zones.

WILCH WENT BACK in at the head of a task force that included pumpers, four-wheel-drive brush trucks, and both structure and wild-land firefighters. They battled traffic as dispatch called them repeatedly to knock on doors where people might still be home. One man walked out his front door and ordered firefighters to surround his home with hoses. The firefighters stuck to their plan and told the man to evacuate.

"I'm staying with my house," he responded, and walked back inside.

"In Mountain Shadows . . . those homes [had] green lawns. They were landscaped. We had asphalt streets. It pretty much seemed to be an area that already had enough defensible space around it," Wilch told me. "But the wood decks that we build off of our houses, the wood chips that we put around our house, the lawn mower that still has fuel that we store under our deck, all of these things are flammable."

"Loser . . . loser . . . loser," firefighters would chant like a mantra as they drove past hopeless homes while looking for ones they had a chance of saving.

STEVE SCHOPPER, A FIREFIGHTER and videographer for the Colorado Springs Fire Department, was amazed to see that the flames had reached Centennial Boulevard, well into the city, and were trying to burn across the four-lane road.

"They gotta make sure that this doesn't get across Flying W Ranch Road," he told the firefighter who was driving as he shot video out the window.

But a house on the other side was already burning. Firefighters from the neighborhood's station were trying to keep the flames from spreading. If it got past that one house, however, it would "take out the rest of the neighborhood to the east of Flying W," Schopper told me. "That was about 277 homes down there."

At a nearby cul-de-sac, a hotshot crew from Redmond, California, had no pump, so they tapped their small hose directly into a hydrant and were spraying a curtain of water between a burning house and the rest of the neighborhood. "They were smoked up and heated up so bad," Schopper said. "You could see the embers bombing down."

Both crews seemed to be fighting losing battles, but an hour later they had managed to keep the fire from spreading out of Mountain Shadows.

"That was the first success," Schopper said. "That was when we turned the corner."

THAT NIGHT COLORADO GOVERNOR JOHN HICKENLOOPER flew over the nine-square-mile burn zone and the firestorm in the northwest corner of the city. "I don't think we've ever seen a wildfire like this in the history of Colorado," he said when he saw the devastation.[9]

"How many people do you think we've lost?" he asked Steve Riker, when they met at the command post after the flight.

"If we haven't lost a dozen people, I'd be surprised," Riker responded.

Jim Schanel was less optimistic.

"I'm guessing we lost 50 to 75 people," he said, but he had difficulty believing his own words. "Seventy-five people, are you kidding me?"

Rich Harvey and Steve Riker, the two incident commanders, hadn't spoken face-to-face during the fire. About 10:30 that night they found themselves standing next to each other.

"Would you like this?" Riker joked, gesturing to the firestorm in the city.

"Absolutely not," Harvey answered. "You're doing a great job. You stay where you're at."

By sunrise 347 homes had burned. But the list of missing persons was going down fast, and all of the firefighters were accounted for.

"Early Thursday morning we started looking at missing people reports," Schanel said. "We were looking at 18 to 25."

In the end, only two residents perished in the blaze. William and Barbara Everett had called relatives to say they were evacuating their home on Rossmere Street in Mountain Shadows before it caught fire. They were found in the home so badly burnt that William was identified by the surgical hardware that held together an old ankle break.

Schanel, Riker, and Harvey winced at the announcement of the fatalities and the final count of destroyed homes, but each had a sad smile later. They had all expected the clearing smoke to reveal many more deaths.

"We dodged a bullet," Schanel said. "We were beyond lucky, and I can't explain it."

22

SEEING RED

Park County, Colorado — June 3, 2012

THE PLANES STARTED FALLING from the sky even before Colorado exploded, but would add another layer of tragedy when the fire season was at its peak. On Sunday, June 3, while hiking through the barren burn scars of the 10-year-old Hayman Fire, I noticed smoke in the distant Rampart Range. The blaze was just a quarter acre in size when the lookout at Devils Head saw it. At a highway overlook I chatted with a ranger who was confident that it wouldn't grow much larger and wouldn't need any aircraft to knock it down. It might have been hard to get planes I learned when I returned to my car and saw a news alert on my phone. Two air tankers had gone down while fighting fires in Utah and Nevada.[1]

One couldn't lower its left landing gear at the Minden-Tahoe Airport, in Nevada, after dropping retardant on a blaze about 50 miles south of Reno. The two pilots kept the plane in the air for 90 minutes to burn off fuel, then brought it in for a belly landing. They scraped down the runway and skidded off the end, leaving a wake of dirt. Both pilots walked away uninjured.

The other's crew wasn't so lucky. Within two hours of the crash landing at the airport, an air tanker swooped low to drop fire retardant on the White Rock Fire, a lightning ignition on the Utah-Nevada border. The plane clipped some trees, a wing hit the ground, and the plane cartwheeled into a rugged canyon filled with piñon pines, junipers, sagebrush, and cheatgrass.

Firefighters hiked for more than an hour to reach the 600-yard debris field. Several fought the fire back from the wreckage while others retrieved the bodies of the pilots, Captain Todd Neal Tompkins and First Officer Ronnie Edwin Chambless. When they retreated, the fire burned over the crash site.

Both planes were P2Vs built in the 1950s primarily to hunt submarines, but with a history of one-way missions. At one point P2Vs were stationed on aircraft carriers and equipped to drop atomic bombs. The planes were incapable of landing on the carriers, so any nuclear mission would require the crew to find a friendly airfield on land or ditch the plane after dropping the bomb.

Korean War surplus, the 60-year-old planes were designed for patrolling over open ocean rather than diving through flaming canyons. A week after the planes went down in Utah and Nevada, Chuck Bushey, the past president of the International Association of Wildland Fire, counted nine fatal crashes of P2Vs, which killed 20 firefighters, since 1987.[2]

The day after the Sunday crashes, authorities grounded the two remaining heavy air tankers available to fight fires in Utah, the latest in a string of groundings and demobilizations of firefighting aircraft.

In 2004 the U.S. Forest Service began removing planes even older than the P2Vs from the fleet after several crashes, some of which involved wings coming off planes. One air tanker crashed in Lyons, Colorado, during the state's epic 2002 wildfire season. A photograph of it streaking toward the ground in flames, with an amputated wing and a fireball spinning behind it, hangs in a hallway by my office, 30 miles south of the crash site. Two pilots died, and two weeks later a helicopter crashed while mopping up the same fire, killing a third airman. At the end of that year, a blue-ribbon panel put together by the chief of the Forest Service reported that "the safety record of fixed-wing aircraft and helicopters used in wildland fire management is unacceptable."[3]

Private contractors operate the majority of the nation's firefighting aircraft, many of which they dig up at the Pentagon's "boneyard" in Arizona to send on the most wing-stressing of missions.

"There are few checks and balances to ensure that the aircraft are airworthy and safe to fly throughout a fire season," the blue-ribbon panel noted. "Contractors have no financial incentive and are not required to ensure that their aircraft are safe to fly . . . Both government employees and contractors assume that Congress, the administration, and federal agencies will never provide the money needed to do the fire-fighting job correctly and safely."

Earlier reports had noted that flight loads and the aggressiveness of firefighting maneuvers exceeded pilot estimates. The current report identified layers of safety concerns, including contracts that prioritized short-term costs over safety, underfunded training, "mission muddle" between land management agencies, and the fact that "the Federal Aviation Administration (FAA) has abrogated any responsibility to ensure the continued airworthiness of 'public-use' aircraft," including military planes retrofitted for firefighting. In addition, according to the report, the FAA "[doesn't] require testing and inspection to ensure that the aircraft are airworthy to perform their intended missions."

When the forest fires of 2012 first ignited, the United States had just 11 heavy air tankers, one-quarter of the 44 available to fight wildfires 10 years earlier. And that was before the two crashes and subsequent groundings.

Demands for the nation to modernize and expand its firefighting fleet came from all sides.

"As the air tanker fleet continues to atrophy, it's going to reduce the country's ability to get there early, which is why so many of these fires mushroom," Democratic senator Ron Wyden, chairman of the Subcommittee on Public Lands and Forests, said the day after the crashes.[4]

Wyden had led a push three months earlier to get the U.S. Forest Service to bring more and newer planes into service. "I have serious concerns about both the size and age of the aging air tanker fleet, and fear that it isn't up to the job of stopping wildfires that grow larger every year."

The public was just as critical of the nation's diminishing firefighting airpower.

Four days before the crashes, Randall Stephens, who blogs about firefighting aircraft, wrote a post titled "Washington Will Let Americans Die in 2012 Wildfires Due to Attention Deficits," a screed attacking Congress and the Obama administration for not providing the Forest Service with $5 billion to purchase next-generation air tankers. (The title of the post has since been changed to "Is Washington Policy

Conducive to a Stable Civil AFF Industry?") Stephens promoted every-thing from purchasing "scoopers," which can drop lake water on fires, to contracting the Supertanker, a 747 converted by the Oregon company Evergreen Aviation to drop 20,000 gallons of retardant.[5] Evergreen has since gone bankrupt, and a new company is operating an updated ver-sion of the 747 Supertanker.

Conservative columnist Michelle Malkin, whose home in the moun-tains above Colorado Springs was threatened by the Waldo Canyon Fire, used the diminishing number of air tankers available to protect her house as an excuse to attack President Obama.[6] A year earlier, Mal-kin pointed out, the Forest Service had canceled a contract with Aero Union, a Sacramento-based company that had supplied eight firefight-ing air tankers but failed to comply with the schedule of safety inspec-tions required by the contract.

"Where there's smoke swirling over Team Obama," Malkin wrote, "there are usually flames of incompetence, cronyism and ideologi-cal zealotry at the source. The ultimate rescue mission? Evacuating Obama's wrecking crew from the White House permanently."

Others recognized that getting rid of the president would have lit-tle impact on wildfires. "Getting into large, multiple wildfire scenarios, there's just not enough [aircraft] to go around," Chuck Bushey told the Associated Press.[7]

At the urging of the Colorado congressional delegation, Obama approved $24 million in funding to buttress the fleet with seven new planes, including next-generation, jet-powered air tankers.[8] Those, along with eight planes leased from California and Canada, would bring the fleet up to 20.

"This is a major milestone in our efforts to modernize the large air tanker fleet," Forest Service chief Tom Tidwell said in a release about the bill.[9]

"It's nice, but this problem isn't fixed with a stroke of the pen," for-mer Forest Service official and bomber pilot Tony Kern told the Denver Post. "You need to have the airplanes available now."[10]

"The USFS should have awarded contracts for at least 20 additional air tankers, not 7," Bill Gabbert opined at Wildfire Today.[11]

In the meantime, the U.S. military would fill the gap. Since the early 1970s Modular Airborne Fire Fighting Systems (MAFFS) have converted C-130 cargo planes into air tankers capable of dropping 3,000 gallons of retardant on wildfires. Eight Air National Guard C-130s equipped to fight fires were already waiting on runways from North Carolina to California.[12] When the Waldo Canyon Fire broke out, Colorado Springs had an ace in the hole, with two of the planes stationed at the city's Peterson Air Force Base.

But federal law prevents the government-owned tankers from fighting fires until all available civilian firefighting planes have been called into service, which points to some of the overarching business tensions in aerial firefighting. After the owners of private air tankers had griped about government planes eating away at their income, federal authorities agreed, in 1975, to mobilize MAFFS only after all suitable commercial resources were committed.[13]

That didn't resolve the argument between public and private aviators, however.

"The bureaucratic tangle and frustration a requesting agency must face to activate MAFFS has served only to distract from the overall value of this unique firefighting system," the Air Force noted in a 1979 report that claimed the tankers were relegated to a "secondary role" and were not widely used, even on the largest and most destructive fires.

In 1978, when a California fire destroyed more than 200 homes, Colonel Russell A. Penland, the air commander for the 146th Airlift Wing, argued that MAFFS could have saved a number of them. The county fire boss had requested large tankers to fight the blaze but had been sent only smaller, commercial aircraft.

Elton Gallegly, a Republican congressman from Simi Valley, California, recalled watching a fire in Camarillo, California, burn over a mountain while two tankers with the California Air National Guard were grounded nearby by the federal restriction. "I'm a private-sector guy," Gallegly said, "but when I sat out on the tarmac watching houses burn in Ventura County and saw the airplanes ready to go and not be able to use them, I got very frustrated."

In the weeks before Colorado blew up in 2012, he filed a bill requiring federal firefighters to use MAFFS as more than just a last resort. The relationship between private contractors, the military, and the Forest Service had more scandalous difficulties in the past.

IN 1987, AFTER A SERIES OF CRASHES, federal authorities permanently grounded the fleet of 1940s-era C-119 Flying Boxcars used by the Forest Service and the Bureau of Land Management to fight wildfires. Late that year Roy D. Reagan, whose company flew some of the C-119s, and Fred Fuchs, the deputy director of fire and aviation for the Forest Service, concocted a scheme. They prodded the U.S. Air Force to list a number of C-130As — the military's most reliable workhorse — as "excess property" and transfer their ownership to the federal government's General Services Administration, which could make them available to other government agencies, so long as those agencies retained ownership of the planes. But instead of government agencies keeping the planes, Fuchs and Reagan secretly used the obscure Historic Aircraft Exchange Program to hand over the newer planes to private contractors in exchange for the grounded C-119s, which were given to aviation museums. The private contractors illegally received title to the government planes in return for commitments to retrofit them to fight wildfires on their own dime and to use the planes only for federal government firefighting missions.[14]

Eventually six contractors received 22 C-130As and six P-3s. Reagan, who served as a broker for the deal, paid himself a commission of four C-130s, which he sold for $1.2 million. His failure to note his commission on his income taxes was an indication, according to prosecutors, that he knew the deals were illegal. The planes given to the private contractors were valued by the Forest Service at as little as $15,000 each during the trade, but an investigation by the U.S. inspector general estimated they were worth about $2.4 million apiece. The aircraft the contractors gave to the government in exchange for them were determined to be largely junk with little historical value.

What's more, the newer planes that were supposed to be dedicated to fighting wildfires in the United States ended up on unrelated missions far from U.S. forests. At least four were stripped, their parts and components sold off. Two were found hauling cargo in Kuwait at the end of the first Gulf War, one was sold to a South African fish hauler, and one, owned by T&G Aviation, crashed in France after a wing came off. Two other T&G C-130s were implicated in drug-smuggling operations in Colombia, Panama, and Mexico.[15]

One pilot, Tosh Plumley, admitted to flying loads of up to 2,500 kilos of cocaine into airfields in California and Arizona for the CIA in C-130s operated by Forest Service contractors.

"All of the contractors had unregulated air fields in remote places that were not usually subject to any kind of Customs inspections," said Gary Eitel, a pilot, aircraft broker, and lawyer who blew the whistle on the scheme and sued the contractors on behalf of the government. "All of the contractors were able to come and go virtually undetected so they could have been doing anything. It was an ideal cover both for drug smuggling and a variety of covert operations."

After the federal Occupational Safety and Health Administration (OSHA) cited the Forest Service for "inadequate use of aviation re-sources" in 1994's South Canyon Fire disaster that killed 14 firefighters on Storm King Mountain outside Glenwood Springs, Colorado, critics noted that many C-130s that were supposed to be available to fight that fire were overseas on missions unrelated to firefighting.

Eitel's muckraking resulted in a Department of Justice investiga-tion and several criminal and civil actions. Fuchs and Reagan served 20 months in prison before their sentences were overturned based on im-proper instructions to their juries. The federal government declined to try them again. A U.S. Forest Service employee who was involved in at-tempts to recover the planes is now the CEO of one of the largest avia-tion contractors for the service.

In 2004, after three of the planes crashed, the U.S. Forest Service permanently grounded all C-130s operated by private contractors, leav-ing only the C-130 MAFFS operated by the military to fight fires.

But the money involved in firefighting aviation continues to fuel tragedies and scandals. At 2008's Iron 44 Fire in Oregon, a helicopter contracted from Carson Helicopters to the U.S. Forest Service clipped a tree while taking off and crashed, killing nine passengers, including seven Forest Service firefighters. It was the worst air disaster involv-ing on-duty firefighters in U.S. history. Investigators determined that the helicopter was overloaded and the manuals guiding its pilots misstated its capabilities. Five years later, a federal grand jury indicted Steven Metheny, the vice president of Carson Helicopters, and Levi Phillips, the company's chief mechanic, for submitting forged performance, weight, and balance information about the helicopters to help the company secure $51 million in Forest Service contracts.[16]

Prosecutors found that Metheny was motivated by "pure greed" in misrepresenting the choppers' weight and how much they could carry. After the accident he schemed with Phillips to conceal their fraud by illegally removing items from helicopters before they were weighed by investigators, manipulating the choppers' fuel gauges to show they were carrying more fuel than they actually were, and falsely claiming to investigators that Carson's scales were out of service. Investigators also discovered that Metheny was stealing and reselling parts and supplies, including helicopter rotors, and using Carson funds to purchase jewelry and pay for renovations to his home.

On June 16, 2015, Judge Ann Aiken, in Oregon's U.S. district court, sentenced Metheny to 12 years and 7 months in prison — a year and a third for every person who died in the crash that resulted from his fraud.

Other revelations documented wildland firefighters' ties to covert military operations overseas. A month after Aiken handed down her verdict, the National Smokejumper Association, the club for firefighters who parachute into wildfires, held its 75th-anniversary celebration in Missoula, Montana. Simultaneously with the event, the summer 2015 issue of *Smokejumper* magazine published the names of 96 smokejumpers who had secretly worked on covert operations for the CIA in Taiwan, Tibet, Congo, Cuba, Vietnam, Laos, and South America.[17] The CIA found that the smokejumpers' skills and equipment for dropping cargo in remote wildernesses, parachuting in themselves, working independently, and keeping their mouths shut were ideal characteristics for covert military intelligence operatives. For instance, after the Chinese invaded Tibet in 1950, smokejumpers worked as "kickers" dropping Tibetan commandos (239 of whom had trained at Camp Hale in Colorado) and pallets of weapons, radios, and supplies into the country. In 1959 those commandos helped spirit the Dalai Lama out of Chinese-occupied Tibet and into India. At CIA headquarters in Langley, Virginia, a painting of the "Khampa Airlift to Tibet" shows guerrillas and equipment being dropped out of a CIA-operated C-130 above Himalayan glaciers 150 miles northeast of Lhasa in the moonlight. The jumpsuits and steerable parachutes in the painting are clearly U.S. smokejumper equipment. Missoula smokejumpers riding in the C-130s oversaw the operation.

U.S. jumpers fought alongside Laotians in their resistance against Laotian and North Vietnamese communists. One Missoula smokejumper, Jerry "Hog" Daniels, who worked for years in Laos and Thailand, is credited with the air evacuation of Hmong resistance leader General Vang Pao and some 2,500 of his soldiers and their families when Laos fell to the communists in 1975. Many of those Hmong ended up in Missoula. Daniels allegedly died of asphyxiation resulting from a leak in his propane water heater in Bangkok in 1982, but his casket was sealed, with orders that it never be opened. Many Hmong claim to have seen him since then.

While the editor of *Smokejumper* magazine, Chuck Sheley, reveres the heroic history of aerial firefighting, even he bristles at the waste in modern firefighting aviation. "I just see all the money wasted with this aircraft business," he told me. After he finished his stint as a jumper, Sheley managed firefighting hand crews for more than 30 years and grew frustrated with how he had to count pennies on the ground while aircraft poured expensive retardant on fires where it would make little difference. "There's no accountability at all for thousands of dollars," he said.

BY LATE JUNE 2012, as wildfires threatened Colorado Springs and Fort Collins, and others erupted all over the West, all the private firefighting aircraft were working blazes, and the U.S. Air Force C-130s joined the action. Less than 24 hours after the Waldo Canyon Fire ignited, aerial porters began loading MAFFS tank systems into the cargo bays of the two C-130s at Peterson Air Force Base. From the runway they could see the fire burning toward Colorado Springs. The planes, with the Air Force Reserve Command's 302nd Airlift Wing, were ready by noon and started flying on Monday, June 25. The next day four of the planes — two based in Colorado and two from Wyoming — ran laps between the base and the fire. With the fire so close, it took just three minutes for the planes to get from the runway to their targets, so both support crews on the ground and the airmen ran like pit crews through their work.[18]

"Having this in town gave us a different sense of urgency about it," Senior Master Sergeant Dave Carey, a C-130 flight engineer on one of the planes who had worked with the firefighting system for 22 years, told *Citizen Airman* magazine. "It was our backyard."[19]

The planes dropped 58,000 gallons of retardant between the fire and the city, but when winds hit 65 miles per hour and the fire took off, the planes were grounded due to smoke and turbulence. Their crews stood on the tarmac and watched the fire roll over the lines of retardant and into the city.

"I'll never forget that feeling of helplessness as we stood on the flight line and saw the excessive winds blowing the fire into the city," Carey said. "There was a blanket of smoke over the whole city. You knew it was coming into town, you just didn't know how far it would go."

The next day the planes were back in the air. They had dropped a total of 133,554 gallons of retardant on the Waldo Canyon Fire by June 30, when they were diverted to fires in Wyoming and South Dakota. Within a day the losses in Colorado Springs would seem small, at least to the Air National Guard.

Edgemont, South Dakota — July 1, 2012

MAFFS 7, A C-130 HERCULES with its six-man crew, arrived in Colorado Springs on June 30 from North Carolina's 145th Airlift Wing, which sent three planes and three dozen airmen to fight the Colorado wildfires. The plane was piloted by Lieutenant Colonel Paul K. Mikeal, who had been on just seven retardant drops as a MAFFS copilot and none as the aircraft's commander. The chief of training for the 156th Airlift Squadron, Major Joseph McCormick, was far more experienced in fighting wildfires with C-130s and was an instructor pilot on the plane. Major Ryan S. David was the plane's navigator, and Master Sergeant Robert Cannon was the flight engineer. Two other airmen rode in the back of the plane to operate the fire retardant delivery equipment.[20]

By the time they arrived, Colorado didn't need them, so the next day, July 1, they made two retardant drops on the Arapaho Fire in the Laramie Mountains of central Wyoming. *MAFFS 7* refilled at the Rocky Mountain Metropolitan Airport near Boulder and was headed back to Wyoming when it was diverted to fires in South Dakota. It refueled again in Cheyenne, then headed to the White Draw Fire, about five miles northeast of Edgemont, South Dakota. A thunderstorm was brewing southeast of the fire when they arrived.

The pilots kept the plane high and watched a small lead plane execute a "show me" run below them. The next time the lead plane swooped over the fire, *MAFFS 7* followed a half mile behind. Its drop was dead on target, but the pilots had trouble maintaining air speed, even gunning the engines to maximum power. Repositioning the flaps from 100 percent to 70 percent got the crew through the run, and they decided to keep the flaps at 70 percent for their second and final drop of retardant.

Five minutes before the air tanker made that pass, Doppler radar detected a very large thunderstorm cell just southwest of the drop zone.

An incident commander with the Army National Guard was riding his motorcycle to the fire's command center and could see the air tanker flying to his right, banking toward the fire. He lost sight of the plane as he descended past a hill, just as he was "hit with this extreme, fierce wind" that blew him to the other side of the highway. A third plane flying about 2,500 feet above the C-130 to manage the air attack was hit by turbulence that bucked the aircraft into 90-degree banks and rapid changes in speed. The pilots could see the fire "sheeting" and smoke lying down across the ground below, indicating "hellacious" surface winds.

At 5:38, as the lead plane and *MAFFS 7* swooped in to make the final retardant drop, the lead plane hit a "bad sinker," falling to within 10 feet of the ground. The pilot "smoked" his engine pulling out of it and would have to go to the Rapid City Regional Airport for repairs.

"I got to go around," the pilot of the lead plane called over his radio, indicating that the planes should circle back and make another run at the target when they were either better aligned with it or when flying conditions were better.

"Yeah, let's go around, out of this," one of the *MAFFS 7* pilots responded.

Thirteen seconds later the lead plane pilot advised the air tanker pilots to "dump your load when you can." Giving up the drop altogether and dumping the retardant would make flying easier.

Three seconds after that the tanker's pilots called, "E-dump, E-dump!" The crew was making an emergency release of the retardant to lighten the plane's load. But while other air tankers can jettison their entire load of retardant, a MAFFS can only spray the slurry in the same way it would drop it on a fire, slowing the effort to lose weight.

A firefighter on the ground saw the tanker dropping its retardant, just as winds she estimated at 50 miles per hour slammed into her.

"Bring some power," one pilot ordered.

"Power's in," the other answered.

"Power!"

"Power's in."

"We're going in!"

"Hold on, crew!"

The massive plane's right wingtip cut through several tree branches. The back end of the fuselage hit the ground. The plane careened up a sparsely wooded slope into a ravine, pivoting right, then rolling left. A propeller chopped the ground, then ripped loose from its engine. The left wing broke off, and another propeller hammered the ground. The number four propeller chopped off the plane's paratrooper door. When *MAFFS 7* came to a stop, the front of the plane lay crushed and folded against the far end of the ravine.

The two crewmen in the back of the plane were severely injured but managed to escape through the door that the propeller had cut open. The plane's emergency locator failed to activate when the plane crashed, but one of them called 911 on his cellphone. Heavy rain, the turbulence of the thunderstorm, and uncertainty about where the plane had gone down kept help from reaching it for about half an hour. When a chopper finally arrived, its crew found one of the survivors still on the phone with 911 and the other wandering, dazed, nearby.

The four crewmen in the cockpit died in the crash, leaving their wives and nine children behind.

MAFFS 7 was the first C-130 equipped with the firefighting system to go down in the 40 years since the technology had been developed. The next day the remaining seven MAFFS were grounded, but with fires spreading through the West and the nation already short of air tankers, they ended up taking only one day off.

THE AIR FORCE INVESTIGATION INTO the crash determined that the microburst that brought down the plane was avoidable. Other planes at the fire had failed to pass on critical information about the thunderstorm and the mission, and the MAFFS crew had underestimated the severity of the conditions. The National Weather Service had issued a severe thunderstorm watch for the area, but there was no evidence that it had been passed on to the crew of the C-130 or that they had requested updated weather information.

"If you add all the pieces up, it was very clear they should not have attempted the second drop," said Brigadier General Randall Guthrie, the Air Force Reserve officer who led the investigation.[21]

In North Carolina flags hung at half-mast and Governor Bev Perdue spoke at the memorial service for the airmen in one of the Air National Guard hangars at Charlotte Douglas International Airport.

In South Dakota residents of Edgemont complained that had firefighters attacked the blaze earlier, it wouldn't have grown to the 9,000 acres at which it was finally contained. Firefighting, particularly with aircraft, is most effective when blazes are young and small. But firefighters elsewhere in the nation wondered what justified the incredible cost, even before the airmen paid the ultimate price, of sending planes into violent weather to fight a fire that threatened only the relics of two historic cabins, some pastures, and forests that were bound to burn someday. In fact, all of the crashes of firefighting aircraft during 2012 — twice the annual average — occurred in remote locations where little private property or infrastructure was at risk.

The twin tragedies of the Waldo Canyon Fire burning right over the more than 125,000 gallons of retardant dropped from planes and, days later, four airmen perishing while flying in stormy weather over rugged ground with little to protect pointed out the contradiction at the heart of America's air war with wildfire.

A firefighter's risk of dying in a wildfire goes up tenfold the moment he or she steps into an aircraft, according to Andy Stahl, executive director at Forest Service Employees for Environmental Ethics. Of the 13 firefighters who perished battling wildfires in 2012, six died in plane crashes.

Yet many firefighters and ecologists question how effective aircraft are in fighting fires.

AT FIRES ACROSS COLORADO I repeatedly stood with homeowners who cheered as swooping air tankers dropped lines of retardant that arced like giant, scarlet Nike Swooshes over the forest. Other times I showed up after a drop to hear that I'd "just missed it," as if it were a performance for which I'd arrived late.

Even a decade ago, some firefighters were calling them "CNN drops" and "air shows" — a bit of theater for news cameras like my own and a public desperate to see something dramatic done to stop the destruction. And some of the most outspoken critics of the nation's air war with wildfire are people who once led the charge — former Forest Service firefighters.

A few weeks before the tankers started crashing in 2012, George Weldon, who was deputy director of fire and aviation for the Forest Service's northern region, based in Montana, when he retired in 2010, filed a Freedom of Information Act request for any data related to the effectiveness of air tankers and what they drop on fires. None of the 30 documents he received dealt with retardant.

"That really underscored the limited success we were having with the aviation program," he told me. "That proved to me that the government doesn't have any scientific studies that deal with the effectiveness of retardant. We have completely taken the large air tanker program out of the context for which it was developed. They're basically a waste of money. The only effect they have is to demonstrate to the public that we're fighting the fire aggressively and doing everything we can."

Since 1995 at least 14 studies conducted by the Forest Service, the Department of the Interior, NASA, the State of Colorado, and several private companies have attempted to determine the effectiveness of firefighting aircraft and retardant in hopes of using that data to determine the number and types of planes and helicopters the nation needs for its firefighting efforts. But a 2013 study by the U.S. Government Accountability Office in response to questions from Congress regarding the use of firefighting aircraft noted that a lack of data gathered about the performance and effectiveness of air power used against wildfires hampered all such research.[22]

For years government reports had asked the U.S. agencies involved in fighting wildfires to collect data on aircraft performance, but not until 2012, the year of the P2V and MAFFS crashes, did the Forest Service begin compiling such data, and then only on large air tankers. The results of that research won't be available for years.

But a 2014 study published in the *International Journal of Wildland Fire* found that in 2010 and 2011, of the fires on which air tankers dropped retardant during the initial attack, nearly 75 percent still escaped containment. And many of the aviation resources were used in extended attacks on longer-burning fires, despite official policies that prioritize their use in initial attacks on new fires, where retardant and water drops are more likely to succeed.[23]

Despite the lack of data on their effectiveness, and their seemingly low success rates, aircraft account for about 25 percent of the Forest Service's firefighting costs, and half of the nation's wildland firefighter fatalities.

To Weldon the use of air power to fight forest fires was an example of industry driving federal wildland firefighting policy. "There's this fire industry that puts more political weight on the agency to spend more money on these fires," he said. "The retardant industry has a powerful lobby in Washington."

"If retardant weren't red, it wouldn't be used anymore," Andy Stahl told me. "There are no studies whatsoever that retardant use saves homes."

All of the various formulas of retardant are basically fertilizer, made primarily from ammonium phosphate and water. Retardant won't extinguish a fire, but it will slow the flames' progress. Firefighters on the ground still have to put it out. Firefighting agencies in the United States have been dropping retardant around fires since the late 1950s. In 2013 the Forest Service dropped 23 million gallons of retardant, which cost about $19 million.

"It's kind of like trying to stop a stampede. You really can't get in front of a stampede and stand your ground," said Cecilia Johnson, a specialist with Wildland Fire Chemical Systems, at the Forest Service's Missoula Technology and Development Center, in Montana. "You turn it. You herd it. You gradually slow it down."

There are lab tests showing the effectiveness of retardant, but field studies would involve so many variables — vegetation types and density; slope angles and aspects; temperature; wind; humidity — that it would be difficult to determine whether the fires were responding to the retardant or other factors.

"A lot's being dropped, and it isn't making any difference," Stahl told me. "But it's terrifically expensive, and it has serious side effects."

Retardant is particularly hard on fish. In 2002 a wayward slurry bomber dropped more than 1,000 gallons of retardant into the Fall River, near Bend, Oregon, wiping out the river's brown trout, redband trout, and whitefish.[24] A lawsuit after the incident forced the Forest Service to remove sodium ferrocyanide from the retardant's formula, and a few months before the tankers crashed in 2012, a judge's ruling forced the Forest Service to keep slurry drops at least 300 feet away from waterways.[25]

But critics say that the air war with wildfire kills far more than fish.

"If we killed ground-based firefighters at the rate we kill aerial firefighters, we'd kill 200 a year," Stahl said, repeating math cited by the 2002 blue-ribbon panel that investigated the safety of firefighting aircraft.

When the Waldo Canyon Fire blew into Colorado Springs, it crossed two wide lines of retardant, he noted. That highlights yet another irony in aerial firefighting. Wildfires are wind-driven events. The conditions that blow up a blaze also make it exceedingly dangerous to fly over them. And as the winds increase, fire retardant and water are less effective, often blowing away before they reach their targets. So when the public and policy makers most want to see aircraft battling a fire is when their efforts are the least effective and the most dangerous.

In less extreme conditions, however, Forest Service officials believe retardant works, and many homeowners and firefighters say they've seen it make a difference. I've seen retardant drops that appeared to slow fires enough for ground crews to stop their spread.

"Actual firefighters of different levels said that's what held this line. That's what allowed them to get the crews in in time. It's making our life a lot easier," said Cecilia Johnson. "I have to believe what they're saying."

THAT DOESN'T MEAN AIRCRAFT ARE USED in the right place at the right time. Often politicians with no firefighting experience push incident commanders to use aircraft they normally wouldn't.

When the U.S. Forest Service's regional aviation chief Ray Quintanar refused to have C-130s drop retardant on California's Cedar Fire in 2003 due to high winds and poor visibility, Representative Duncan L. Hunter, a Republican from San Diego County whose house was threatened by the fire, called his friend General Richard B. Myers, the chairman of the Joint Chiefs of Staff in Washington, D.C. Myers arranged for military tankers to fly from North Carolina, Wyoming, and Colorado to the California blaze, despite Quintanar's insistence that they wouldn't help. When the planes arrived, the weather grounded them. Hunter's house burned. Yet he remained unapologetic about going over the chief's head and noted that the military tankers fought other fires in California during their western swing.[26]

And, of course, the stress incident commanders endure when being pressured to put planes in the air pales in comparison to the grief and guilt they bear when one of their aircraft goes down.

A MONTH AFTER THE MAFFS CRASH in Edgemont, South Dakota, Anne Veseth, a 20-year-old seasonal wildland firefighter, was excited to take her first helicopter ride—a short jaunt to the Black Mountain Lookout in the Clearwater National Forest, about 100 miles from her home in Moscow, Idaho.[27] She landed safely, but the fire business brings death from the sky in other forms.

A week later I stood with nearly 300 U.S. government firefighters watching bagpipers and an honor guard lead her family into the Church of the Nazarene in Moscow. The colors of mourning were the yellow and green hues of the firefighters' clothes, although some were so coated with soot and ash that they looked gray.

Veseth's family urged firefighters to dress for her funeral the same way they would if they were showing up to fight a fire with Anne.

At her funeral, eyes accustomed to tearing up from smoke and sweat wept. Many also seethed with anger, after seeing a report from a hotshot crew that had refused to engage in the firefight the day before Veseth's death because they deemed the operation "extremely unsafe" and had warned the leaders of the firefight of the very thing that killed her. The leaders of the fight against the Steep Corner Fire, however, weren't part of the federal firefighting effort, but instead were fighting the blaze on behalf of the timber companies that owned the land where it had started.

While Veseth was enjoying her chopper ride, loggers about five and a half miles from the lookout she was resupplying noticed a fire in slash from their timber harvest.[28] The site, 56 miles northeast of Orofino, Idaho, belonged to the Potlatch Corporation, a timber company, but it was just above national forest land dense with conifers. The loggers contacted the Clearwater-Potlatch Timber Protective Association, a private group that puts out fires in logging company forests. But because the fire threatened public land, Veseth's crew — Engine 31 of the Clearwater National Forest's North Fork Ranger District fire crew — was also assigned. Her crewmates picked her up when her helicopter landed.

It's unusual, but not unheard-of, for federal firefighters to work fires on private timber operations. The last time government firefighters worked on a C-PTPA fire had been 16 years earlier. Over the previous two decades, the C-PTPA had fought about 1,850 fires, but only about 1,600 acres had burned, a reflection of its aggressive initial attacks to protect valuable timber.

Veseth's crew arrived around four in the afternoon and moved fast through the uncut forest on the southern flank of the fire, but slowed when they got to felled timber up to eight feet deep. That, and slopes so steep the loggers had to use cables to haul out logs, prompted the incident commander to keep the firefighters off the line in the dark.

In the morning the leaders ordered Veseth's crew to move past areas threatened by teetering trees, leaving gaps in the fire line. Her crew's engine boss complained to the division supervisor about the gaps: digging an "unanchored" fire line violates basic firefighting safety training.

Montana's Flathead Hotshots arrived at the fire at 2 p.m., and their supervisor and foreman hiked different sides of the fire line. The hotshots noticed that many of the firefighters on the scene, including the commander, were wearing jeans rather than the fire-retardant clothing government rules require. Some weren't carrying fire shelters, another required piece of personal protective equipment (PPE). Others were running chainsaws without safety gear.[29]

The hotshots' report, which they posted on SAFENET, a website where wildland firefighters can report safety concerns, describes a "hodge-podge" of firefighters isolated by poor communication, weak direction, and areas deemed too dangerous to work due to falling trees and flames. There was no site for evacuating injured firefighters and no medical plan.[30]

A fire crew made up of prison inmates was repeatedly chased uphill by the fire and forced to dodge trees and boulders that rolled down on them from above. "A huge snag came down above us and started rolling down through the standing trees," the report states. "The prison [crew boss] commented, 'That is the sound of the day.'"

The hotshots repeatedly requested that helicopters drop water on the fire threatening the prison crew, but to no avail, while elsewhere water was dropped on firefighters rather than flames.

"We had heard multiple people asking, 'are you ok' after helicopter drops," the report notes. "The people directing helicopter drops had no or little experience utilizing helicopters and were having the helicopters drop water without clearing the line of personnel."

One sawyer shut down his saw and asked himself, "What are we doing here?" after a helicopter nearly hit his crew with a water drop. When the prison crew finally withdrew from the fire-threatened drainage, injured firefighters slowed their retreat.

The hotshots noted violations of 8 of the 10 basic safety rules for federal wildland firefighters. In addition, 12 of the 18 Watchout Situations that firefighters are required to mitigate went ignored.

The incident commander requested that the hotshots fill in the gaps in the fire line that were deemed unsafe for the other firefighters, but the hotshots wouldn't engage the fire until their safety issues were addressed and gave him a written list of their concerns. Chief among them was the number of dead and fire-weakened trees — snags — that were falling around firefighters.

The commander responded that his team "had a different set of values and do things differently," according to the hotshots' report.

"We're doing the best we can with what we've got," another C-PTPA firefighter told them.

The hotshots declined the assignment and left.

According to the Serious Accident Investigation Report filed six months later, leaders on the fire attempted to mitigate the hazards.[31] They put together an Incident Action Plan with two objectives — ensuring firefighter safety and minimizing the loss of timber resources.

Veseth's crew also disengaged from the fire that night due to safety concerns, but they returned the following day after leaders assured them that the hazards would be taken care of. More crews were on the way, they were told, and two professional tree fallers were dropping hazardous trees. "All the things we wanted," a crew member told investigators.

They stopped for lunch along Steep Creek. Veseth was watching one of the tree fallers cut down a snag when a colleague standing with her saw another tree coming down toward them from across the creek.

"Snag falling!" he shouted.

"Everyone scatters," the Serious Accident Investigation Report recounts. "Some firefighters run downhill while Kerry [an alias used in the report] and Anne run uphill."

The falling snag knocked over another tree, like a domino.

"Down!" the other firefighter yelled, turning to run downhill away from the second falling tree.

"With his fists and teeth clenched, he expects to be hit. He hears a tremendous sound as the trees crash downward and feels the whip of limbs on each side of him. He falls down but, upon realizing he is uninjured, quickly gets up and looks for Anne who he thought had been right behind him. He finds her three or four strides uphill under the tree branches. After quickly clearing them away, he determines she did not survive."

Where it hit her, the cedar was 13 inches in diameter and 123 feet from the stump it had broken off of.

WHEN VESETH DIED, the Forest Service's $948 million firefighting budget was nearly exhausted, but the fire season still had months to go and was projected to cost as much as $1.4 billion.

The National Cohesive Wildland Fire Management Strategy states that "safe aggressive initial attack is often the best suppression strategy to keep unwanted wildfires small and costs down." But that philosophy, regardless of how safe the initial attack is, puts more people in harm's way. Most fire line deaths occur in the early "initial attack" or "transition" phases of firefighting operations, when teams or individuals may take on blazes without adequate management, communication, or knowledge of the terrain and weather.[32] Initial attacks are often made up of firefighters from a variety of jurisdictions who can prove difficult to coordinate and may have differing philosophies of firefighting.

In his memo three months before Veseth's death, James E. Hubbard, the Forest Service's deputy chief for state and private forestry, said the service's dwindling budget required fast attacks on even the most remote fires to save money. Many firefighters, however, questioned the cost savings of putting out every fire in extreme fire years. "It's externalizing long-term costs, including the highest cost — firefighters' lives," said Tim Ingalsbee, executive director of the watchdog group Firefighters United for Safety, Ethics, and Ecology.

After Veseth's funeral, at a meeting of the Idaho Land Board, state attorney general Lawrence Wasden grilled the Idaho state forester, David Groeschl, about firefighter safety.[33] Yet despite the SAFENET list of safety violations, the Serious Accident Investigation Report concluded that the "judgments and decisions of the firefighters involved in the Steep Corner Fire were appropriate" and that there were no "reckless actions or violations of policy or protocol."[34]

The Occupational Safety and Health Administration issued citations against the C-PTPA and fined it $10,500 for the safety violations noted in the SAFENET report.[35] That OSHA issued citations and fines for safety violations and broken rules in fighting the fire that killed Anne

Veseth, while the Serious Accident Investigation Report downplayed them, foreshadowed the investigations into the tragedy on Yarnell Hill in Arizona a year later. But rather than one fatality for which nobody was assigned blame, there would be 19.

VESETH'S FAMILY RESERVED JUDGMENT on the circumstances of her death. That wasn't surprising, given the family's devotion to public service. Veseth's mother, Claire, was a nurse; her oldest sister, Rachel Tiegs, a paramedic; and her brother, Brian, a firefighter.

In 2010 Anne approached her brother about getting a job as a wildland firefighter.[36] The previous year she had completed an auto mechanics degree at Lewis-Clark State College, although she had never shown an interest in working on cars before enrolling in the program. That same year she was second runner-up in the Moscow Junior Miss competition.

"She picked breakdancing as her talent and mom informed her that she didn't know how to breakdance," her brother, Brian, told the crowd at her funeral. "But she took lessons and she nailed it."

She was in her second season with the North Fork Ranger District fire crew in the Clearwater National Forest, just a couple of hours' drive from her home, and had eagerly signed on with teams sent to fight fires in Colorado (the High Park Fire) and Arizona earlier in the year.

When she arrived at the Steep Corner Fire, she was 10 days away from starting another degree program, this one focusing on forestry and fire ecology. She finally seemed to have found her path in life.

Outside her funeral, after the bagpipes and bells, Anne Veseth's family walked through two rows of firefighters, climbed into her brother's fire truck, and slowly drove around the church. Firefighters lined the road, then climbed into their trucks to join the short procession.

With 95-degree temperatures driving scores of blazes in Idaho, Washington, and California, some of those who hadn't washed their firefighting clothes for the funeral would not be changing out of them before they were back on a fire line.

23

NEVER WINTER

Rocky Mountain National Park — October 9, 2012

BY THE TIME OF ANNE VESETH'S DEATH in August, fires were on track to burn a record amount of U.S. land in 2012, surpassing the 9.8 million acres burned in 2006. The nearly 7 million acres that had burned by the middle of the month was 22 percent higher than the 10-year average, but the number of fires reported was 25 percent below the average. The disparity showed that the fires were getting bigger.[1]

Evidence in tree rings and lake sediments indicates that there were actually seasons with bigger fires that burned even more western acres before U.S. agencies started keeping such records. But, of course, that was also before the expanding nation cut, paved, and developed much of the West, and before it started spending billions of dollars trying to eradicate wildfires.

During the Hayman Fire a decade earlier, Colorado governor Bill Owens drew heat for saying that it seemed like "all of Colorado [was] on fire." In 2012 it seemed like the whole nation was burning.

In Idaho, where Anne Veseth lived and died, National Guard troops joined the fight against at least nine large fires burning in the state during the week after her accident. One burned over 100 square miles and threatened to overrun Pine and Featherville. Another stranded 250 rafters for days on the Middle Fork of the Salmon River. A few weeks later a wildfire burned over closed uranium mines, threatening to add radioactivity to the fire's smoke.[2]

With parching drought and 113-degree temperatures not seen since the 1930s Dust Bowl, Oklahoma faced at least 11 large wildfires in a single day. Residents grabbed their children and ran as fire raced through their neighbors' yards, destroying 25 homes near Norman and 40 outside Tulsa. Authorities were hunting for a man seen throwing flaming newspapers from his pickup truck onto the grass.[3] A football player for Emporia State University missed the first day of training after surviving one of the Oklahoma fires by driving his car into a farm pond.[4]

In central Washington's Kittitas County, a crew barely outran a wind-driven blowup as they drove out of the Taylor Bridge Fire. The fire, about 75 miles east of Seattle, burned 22,000 acres and some 60 homes. The fire came within 100 feet of the Chimpanzee Sanctuary Northwest, near Cle Elum, but the seven chimps there survived.[5]

In Utah a lightning-sparked fire threatened a herd of wild horses and closed the Pony Express Road in the state's western desert.[6] In California some 8,000 firefighters battled nearly a dozen large fires. One 66-square-mile blaze in the Plumas National Forest threatened a clothing-optional resort and 900 homes. A brutal heat wave drove flames through more than 24 square miles of brush in a day.[7]

Three days after Anne Veseth died, a jury in San Bernardino, California, convicted Rickie Lee Fowler of five counts of murder for the fatal heart attacks that occurred during the Old Fire, which destroyed more than 1,000 buildings in 2003 and which prosecutors claimed Fowler intentionally set in dry grass. The judge sentenced Fowler to death. The second sentence of capital punishment for a wildland arsonist in U.S. history, it came just three years after the first.[8]

Meteorologists warned that drought would extend the western fire season by two weeks. They underestimated.

Colorado finished the year as it had started it, with a wildfire burning not just a couple of weeks outside the normal fire season, but in winter. Unlike the Heartstrong Fire, which the previous winter had raced across the plains and died out within hours of igniting, this one would burn from autumn until spring among the high peaks of Rocky Mountain National Park.

"If you called me up and said, 'Hey, we've got a fire. You guess where it is,' it's the last place I would have guessed," Jason Sibold, a wildfire researcher at Colorado State University, said in an interview.

An illegal campfire on October 9 ignited Forest Canyon, a remote gorge filled with trees and dead timber that hadn't burned in some 800 years. November stayed warm with little precipitation. When the snow did come, the ground was cold enough that it didn't melt to wet down the fuels.

On Thanksgiving, Mike Lewelling, the fire management officer for the park, found the blaze burning under a foot of snow. Early on the morning of December 1, winds gusting to 70 miles per hour pushed the fire more than three miles in 35 minutes, forcing evacuations of more than 1,500 people from homes around the park and burning down a cabin. Much of the terrain where the fire was burning was so steep and dense with fallen trees that firefighters couldn't confront the blaze there.[9] Even if they could, there were hardly enough of them available.

The Alpine Hotshots, based in the park, had been furloughed three days before the fire started, having racked up about 1,000 hours of overtime. Nonetheless, those still around got on the blaze fast. Some of their families were among the evacuees. Other local wildland firefighters were on mandatory days off after returning from fires out of state, but they joined rangers and trail crews to evacuate the park. Firefighters from other states arrived as soon as they could.

Almost all of the other firefighting resources in the region were demobilized for the winter. It was yet another challenge the expanding fire season presented to the country. During the winter wildfire season, college students, who made up a substantial portion of the seasonal firefighters, weren't available. Contractors had most of their helicopters, planes, and bulldozers in storage. Keeping them available for freak winter fires would add millions to the nation's firefighting bill. The short, cold days meant that the firefighters stayed in hotels rather than tents, adding even more expense. Helicopters that were available to drop water on the blaze found the lakes they dipped their buckets into freezing over.

Rich Harvey was once again called in from Nevada and his Type 1 Incident Management Team returned to Colorado to work into the winter.

Ron Wakimoto, a fire scientist at the University of Montana, once told me that when white settlers first came to the Rocky Mountains, in the early spring they saw native fires on high mountains still thick with snow and wondered how the Indians got up there to set them ablaze. In fact, they learned later, the natives set logs afire just before the winter snows arrived. Their coals would smolder through the winter beneath the heaviest snows, then reignite as the spring warmed and the snow melted away. The resurgent fires improved the wildlife habitat in their summer hunting grounds and signaled it was time to head for the mountains.

When snow blanketed Rocky Mountain National Park in the winter of 2012–2013, the Fern Lake Fire did the same thing as the Indians' fires, burning beneath the heavy, frozen blanket in anticipation of running free in the spring.

Firefighters, land managers, and scientists saw the blaze as both an exclamation point on a Colorado fire season that had been far more destructive than any on record and a harbinger of things to come. "It's downright shocking that we are dealing with fire in January," said Jim White, director of the Institute of Arctic and Alpine Research at the University of Colorado.

It wasn't just the warming or the drying climate, but the fact that the fuel was warm and dry for a much longer portion of the year, explained Tom Veblen, of the university's Biogeography Lab.

"Fire season used to run from late May through September," said Scott Dorman, the chief of the Estes Valley Fire Protection District, which fought the fire in Rocky Mountain National Park in the dead of winter six months after battling the Woodland Heights Fire in the heat of the summer. "We're finding out that there isn't such a thing as a fire season anymore. It's kind of year-round."

"Wherever we look, we keep coming back to climate, climate, climate," Jason Sibold said. "The next 10 years, 20 years, 30 years, are projected to be more of the same, if not more intense."

Even with blazes burning right into the new year, 2012 didn't break the record for acreage burned. But the year did set plenty of other records, including for the Fern Lake Fire, which not only burned in winter among some of the highest mountains in the contiguous United States, but wasn't out until more than six months after it ignited. So, in the Rocky Mountains, the 2012 fire season was most notable for the fact that it didn't end, but burned seamlessly into 2013.

24

BLACK FOREST

Black Forest, Colorado — June 11, 2013

THE FERN LAKE FIRE THAT BURNED through the winter in Rocky Mountain National Park showed no smoke after January, but it wasn't declared extinguished until June 25, 2013. By then the rest of Colorado was burning again.

If the Lower North Fork Fire set the direction for the 2012 conflagrations, the following year covered the remaining points of Colorado's compass of conflagration. The East Peak Fire near Walsenburg destroyed 14 homes, and the East Fork Fire overran the spruce and fir forests of the San Juan Mountains. Lightning strikes near Wolf Creek Pass, also in the San Juans, ignited the West Fork, Windy Pass, and Papoose Fires, which converged and spread over 110 square miles. The combined fire was the fastest and hottest on record in the Rio Grande National Forest, which was so ravaged by spruce beetles that few of its standing spruce trees were alive.[1]

"It's like gasoline up there," Cindy Shank, a former firefighter and executive director of the southwest Colorado chapter of the Red Cross, told the Associated Press. "I've never seen a fire do this before. It's really extreme, extreme fire behavior."[2]

The fire threatened the Wolf Creek Ski Area, spread south to force the evacuation of more than 1,000 people from the towns of South Fork and Del Norte, and finally turned north into the historic mining town of Creede. A smoke plume rose 30,000 feet into the sky as the blaze came within a few miles of the towns. Thousands of firefighters with tankers, brush trucks, and two very large air tankers, or VLATs — DC-10 jets set up to drop 11,600 gallons of retardant — massed between the blaze and South Fork, but the fire never made it to the town.

Catastrophe instead returned to Colorado Springs.

AFTER THE WALDO CANYON FIRE, officials from Colorado Springs, the state, and the FBI spent months investigating its suspicious start. But aside from determining that it was definitely "human caused," they couldn't say whether it was arson or an accident. Unless a witness came forward, they would probably never know for sure.[3]

On June 11, days shy of the one-year anniversary of that fire's start, another suspicious fire blew up outside Colorado Springs. This one didn't come down from the mountains, but ignited on the plains just northeast of the city, in Black Forest. The ponderosa pine woods there were badly overgrown due to decades of fire suppression and were populated with residents who often resisted any rules, restrictions, or suggestions about how to care for their property.

"Both residents and firefighters, we've talked about Black Forest having 'the Big One' since I've been here," Jim Schanel, Colorado Springs' most experienced wildland firefighter, said. "It's just an unhealthy forest with a lot of people in it, and with poor access and egress."

As opposed to the days-long wait for the Waldo Canyon Fire to burn into the homes of Colorado Springs a year earlier, the first flames reported from the Black Forest Fire were coming from a house or shed in the forest.[4] In the following 36 hours it burned 488 homes, making it the most destructive fire in Colorado history — the second fire outside Colorado Springs to set that record in a year and the fourth in the state to break it in four years.

THE DAY OF THE FIRE, a Red Flag Warning predicted temperatures above 90 degrees, low humidity, and winds gusting to 45 miles per hour. Shortly before 1 p.m. another suspicious fire broke out in Royal Gorge Bridge and Park, 65 miles from Colorado Springs. That blaze forced the evacuation of hundreds of tourists and staff, destroyed 50 buildings, and leapt across the deep gorge carved by the Arkansas River.

When Black Forest fire chief Bob Harvey (unrelated to Rich Harvey, the Type 1 incident commander who would eventually take command of the fire) first responded to reports of smoke, he wondered if it was coming from the clearly visible plume rising from Royal Gorge. But at 1:42 a control tower at the U.S. Air Force Academy reported smoke rising east of the New Life Church in Black Forest. Other calls reported a grass fire that was already spreading into the trees. A resident, Gregg Cawlfield, made a cellphone video of a burning outbuilding near the intersection of State Highway 83 and Shoup Road. But though white smoke spread through the woods, the flames there were initially hard to find.[5]

Engines and brush trucks from at least six departments responded, without knowing which jurisdiction the fire was in. The winds were calm when Ric Smith, a Black Forest volunteer with 28 years of wildland firefighting experience, found the fire burning in a heavily vegetated bowl along Falcon Drive. It was creeping along the ground, and in the light wind Smith was confident he could "hook" the fire before it spread to any houses or farther into the forest. He laid about 650 feet of hose along a ridgeline, but at 2:18 the winds suddenly picked up to 30 miles per hour. Firefighters "heard someone yelling over the radio to evacuate, evacuate, evacuate," as a "wheel of fire" blew into the treetops with a boom. It chased Smith back to his truck, but not before one radio call reported that he was a "goner."[6]

Houses ignited as the firefighters retreated. One of them reported propane tanks exploding like popcorn. The temperature hit a record 97 degrees, and the winds blew the blaze to more than 100 acres in minutes. The flames jumped fire lines and ignited multiple spot fires.[7]

Without the slopes of the Rockies, which tended to pull the Waldo Canyon Fire uphill and away from the city, the Black Forest Fire progressed even more erratically. Fire fronts pushed north, east, and south. Flames as high as 300 feet burned homes scattered through the woods. The fire would stretch out in one direction, then turn with the winds into a wide, flaming front. Hot spots blew up almost randomly to destroy neighborhoods, and spot fires ignited up to a mile from the blaze's many fronts. Evacuations spread over three counties and nearly 150 square miles, removing some 38,000 people from 13,000 homes.

Marc Herklotz and his wife, Robin, who both worked for the Air Force Space Command, were sitting on their porch watching television while their neighbors, Bob and Barb Schmidt, were racing to save what few items they could after getting a call to evacuate. When Barb departed, she urged the Herklotzes to flee, but they said that they hadn't received a reverse 911 call and that they'd know when to go. At about 4:20 the couple told a friend over the phone that they could see the glow of the fire and were packing to evacuate.

"At 5 o'clock there's another phone conversation," El Paso County sheriff Terry Maketa said. "The person that [the Herklotzes] were speaking with said he could hear popping and cracking in the background, and they advised that they were leaving right now."

The Herklotzes' bodies were found beside their car in the collapsed garage of their burnt home.

"I don't know if it matters," Maketa responded when neighbors and reporters asked about the lack of a call to evacuate. "They saw the fire. They admitted they saw it. They saw it coming, and I don't think we need phone calls to tell us when it's time to go. We can all look up and see orange and fire and know it's time to leave."[8]

The Herklotzes' deaths were just one of the controversies Maketa would find himself embroiled in.

Chief Bob Harvey came to the fire in an unmarked Chevy Suburban with no radio or emergency equipment. As the initial incident commander, he tried to manage the firefighting effort by switching between the channels on a single handheld radio, but the airwaves were overwhelmed with traffic. Firefighters couldn't reach him, and he missed key transmissions. As resources flooded the staging area, he had only one map book to organize them.[9]

Scott Campbell, as the county's assistant deputy fire marshal, was part of Maketa's team. He ordered helicopters and two heavy air tankers as he was on the way to the fire, but he was told that at least one of the helicopters was already assigned to the Royal Gorge Fire. They'd let him know later about the other aircraft.

Right after the fire blew up, Campbell asked to take over command of the incident from Bob Harvey — as the fire grew, standard protocol would have moved it up from local to county and finally to federal control — but Harvey told him it was a bad time to transfer authority and put him off. He transferred the fire to the county's authority sometime within the next 90 minutes, but afterward there were only a couple of handwritten notes to prove it. It was the beginning of a series of escalating disputes between Sheriff Maketa and the fire chief.

Those tensions started with Campbell and Harvey, who were less than friendly, but whom Sheriff Maketa told to work together. Instead, Harvey and his men reported being shunned. They were removed from management of the fire, and some members of the new incident command wouldn't speak to them, preventing them from sharing notes about the fire and the resources available to fight it. Harvey and the other officer who had been managing the blaze left the command post and went "freelancing" — making up their own crew to fight the blaze independent of incident command.[10]

By seven that night two helicopters were on the fire. Military planes would start bombing the blaze in the morning. The Black Forest Fire had more aircraft fighting it than even the Waldo Canyon Fire had. But with yet another Red Flag Warning, the fire would remain at 0 percent containment for days.

On Thursday Rich Harvey, the incident commander with the federal team that had fought the Waldo Canyon, Lower North Fork, Fern Lake, and Fourmile Canyon Fires, arrived from Idaho to lead the fight against the Black Forest Fire. His trips to Colorado were turning into reunions.

Jim Schanel watched wisps of smoke over Black Forest turn into a plume as he was driving home from an emergency management class in Colorado Springs. Harvey brought him on as a structure protection specialist. Steve Riker, the Colorado Springs incident commander with whom Harvey had worked at the Waldo Canyon Fire, led a task force here as well. He was less than a mile from his home in Monument when the fire forced his team to retreat.

Riker called his wife to tell her to get ready to evacuate. "If the wind would have kept at that 65 miles an hour, I have absolutely no doubt that Monument would not have been safe, 10, 12 miles up the road," he told me. "We would have burned that fire probably to Castle Rock before we got control of it."

But the winds did die down, the temperature cooled, and it started to rain. Ten days after it began, the Black Forest Fire was fully contained.

The firestorm between the men who had led the initial attack, however, would continue for more than a year.

Shortly after the fire Bob Harvey told NewsChannel 13 KRDO in Colorado Springs that evidence indicated the fire was intentionally set. Maketa snapped back that while the fire appeared to be human caused, investigators found nothing to indicate it was arson and attacked Harvey for botching the initial response.

Chief Harvey then filed a report on SAFENET, claiming that around 2:30 on the morning after the fire ignited, Scott Campbell, the assistant deputy fire marshal who took over the fire, started a burn using a drip torch to remove excess grass and brush.[11] The chief noted that there was no safety plan or briefing about the burnout, no communication about it, and nobody left to monitor it. He pointed out that it was a violation of protocol for an incident commander to leave the command post to man a torch or hose. Harvey warned a crew in the path of the burn to get out of the way. Afterward, several mobile homes in the area burned, but nobody can say whether it was the flames from the backburn or the wildfire that destroyed them.

Nine months after the fire, the Black Forest Fire/Rescue Protection District released a 2,000-page report exonerating Bob Harvey for the mistakes made during the initial attack of the fire and making its own accusations against Campbell and Maketa. Most damning was the allegation that a team of firefighters who had nearly been burned over on the first night of the fire were sent, along with "a very key piece of equipment from Falcon Fire, their 2,000-gallon 'tactical tender,'" on a "secret special assignment" to protect a single home. The home's address and owner were not to be aired "over any radio channel," but the property was later revealed to belong to Bob McDonald, the acting commander of emergency services for El Paso County. Steve Thyme, a member of the Colorado Springs Fire Department and the El Paso County Wildland Fire Crew, said the order "came from the top," which, the report said, was incident commander Scott Campbell. McDonald was Campbell's boss.[12]

The property's exposure to the fire hazard was well mitigated by defensible space cut into the woods surrounding the home, and it had its sprinklers on, so the firefighters never had to put any water on it, but a neighboring house burned down. "Protecting a public official's home at the expense of other residences is poor decision-making at best," the report said.

The morning following the firefighters' efforts, witnesses saw McDonald visit his home, where he hugged the firefighters. Maketa also visited the house, the report said, riding in an SUV with two females. Maketa responded that the firefighters were there to protect multiple homes and the only woman with him when he visited the scene was his wife.

In a 50-minute press conference, Maketa called the report "garbage, slanderous," and said "it needs to be burned."[13] He also lashed out at the report's speculation that there might be a connection between the human starts of the Waldo Canyon Fire and the Black Forest Fire.

Two months later Maketa released his own report, which was less than 10 percent as long as the previous report and far less incendiary, but made clear that the sheriff blamed the Black Forest fire chief for the chaotic response to the blaze.[14]

25

TRICKLE DOWN

Colorado Springs — August 9, 2013

THE COLORADO SPRINGS' FIRES DESTRUCTION came in the form of water as well as flames and finger-pointing.

Two months after the Black Forest Fire, Laura Hunt, a U.S. Army veteran, was in her small cottage on Canyon Avenue, next to a culvert off Williams Creek in Manitou Springs. "I'm sitting there, and all of a sudden water starts pouring into my living room window," she told me. "I had no time to gather anything, not my pets or anything."

She ran out her front door and headed across the street to higher ground. "As I was doing that, I got washed away. I was underwater and flailing."

The water was flowing at 30 miles per hour. She grabbed a tree, then pulled herself onto an embankment and crawled above the flood. "Then I noticed I was walking on a broken leg and a broken foot," she said.

Neighbors carried her away from the creek and washed the mud and cinders out of her eyes. Her street was full of boulders, mud, and rubble.

"It wasn't until the next day that I found out that my cottage totally is gone," Laura said. "There's just a slab there. My cats, when they get scared, they hide in a drawer underneath my bed. So that's probably what they did. I lost everything."

It was the third flood to hit Manitou Springs that year, and the fourth to inundate Colorado Springs since the Waldo Canyon Fire.

Wildfires often not only leave mountain slopes denuded of vegetation that would hold them in place, but also turn the soils hydrophobic.

Oils from the burning vegetation spread across the dirt below, making it impervious to moisture. Water runs fast down the water-repellent soils on barren, burnt slopes. The velocity of the flow carries away materials that normally would be as solid as rock — tons of mud, charred branches, and trees, as well as boulders, cars, and even houses. Floods on burnt mountainsides can carry 10 times more debris than those on slopes that aren't burnt.[1]

"I saw a guy running in a trickle of water," said Chris Johnson, who lives in a castle-like house near the Cave of the Winds, where Williams Canyon enters Manitou Springs. "He was just screaming . . . Behind him was a wall of water."

Half a dozen people on foot barely outran the flood. The debris flow swept a pickup truck down Canyon Avenue, and then a small car. Laura Hunt's yellow cottage bobbed down the street on the floodwaters, then fell apart, leaving one wall lying across a sidewalk. Other homes broke in half, filled with mud, or had appliance-sized boulders deposited in the living room. Days later a Bobcat earthmoving machine that had been washed away still hadn't been found. Rock and mud filled the shops throughout Manitou's tourist district.

While watching coverage on TV, Johnson saw his '73 Chevy truck, which his grandfather had bought off the line, on top of his next-door neighbor's '69 BMW in floodwaters a half mile from their homes. On U.S. 24, which runs into the mountains above Colorado Springs, a man was found buried in rubble after being washed away from his car or trying to outrun the flood, while a woman was rescued from a tree above the floodwaters.[2]

The Hayman Fire forced Johnson from a cabin he lived in. Ten years later he evacuated his current home in the Waldo Canyon Fire. "But nothing compares to watching that flood come down and the people running away," he said.

Post-wildfire debris flows buried roads, clogged water systems, and blackened rivers across Colorado during the summer of 2013, leading up to a "biblical" flood that washed out highways and homes and killed five people that September.

Mudslides and debris flows are often the insults that follow the injury of wildfires. On Labor Day 1994 heavy rains drove a debris flow from Storm King Mountain, where 14 firefighters had perished in the South Canyon Fire two months earlier.[3] The slide buried Interstate 70 and washed 30 cars into the Colorado River.

In California, debris flows after wildfires have buried neighborhoods up to their roofs. On Christmas Eve 2003, four inches of rain fell in 24 hours on the steep, burnt slopes left by the state's biggest wildfire, which had burned during the previous summer and fall. A dozen debris flows broke loose on Christmas Day, killing three people, burying a church camp, and bulldozing through the town of Devore.[4] A 1934 postfire landslide killed 49 people when a 20-foot-high wall of rock inundated La Crescenta–Montrose.

But in 2013, a few weeks after the Black Forest Fire, California learned that the impacts of a wildfire trickle down to communities far from the mountains' flames and floods.

ON AUGUST 16 A HUNTER'S ILLEGAL CAMPFIRE escaped near the Rim of the World overlook in the Stanislaus National Forest outside Yosemite National Park. Record drought had cut 10 inches from the region's average precipitation for the year. The Sierra Nevada held near record-low snowpacks.

"The situation is extremely crispy and dry," Bill Patzert, a climatologist with NASA's Jet Propulsion Laboratory, told the *Los Angeles Times* earlier in the summer. "That equals incendiary."[5]

The drought and a heat wave primed the forest to burn, and a century of putting out fires had left much of it overgrown, with plenty of fuel.[6]

The Rim Fire blew up to 100,000 acres in 36 hours, and engulfed nearly 90,000 acres more in the following two days. When it was finally contained more than two months later, it had burned 257,314 acres, making it the third-largest wildfire in California history and the largest on record in the Sierra Nevada. Nearly 80,000 of the acres burned were in Yosemite National Park. The fire burned so intensely that firefighters cleared brush around giant sequoia trees, which are normally insulated from wildfire by their thick bark, and surrounded them with sprinklers to keep them from igniting. Some wood smoldered right through the winter, and the fire wasn't declared out for more than a year after it was contained.

Its impacts ran down to cities hundreds of miles away.

The fire burned to within a mile of the Hetch Hetchy Reservoir, which supplies water to more than 85 percent of San Francisco's population. While the water was never so contaminated with ash that the city's supply had to be cut off, as many residents had feared, it forced the closure of two hydroelectric power stations that depend on the reservoir. The San Francisco Public Utilities Commission purchased $600,000 worth of electricity from other sources to prevent blackouts.[7] In the end, the fire cost the city $36.3 million in repairs to infrastructure, power purchases, and remediation of the landscape to prevent post-fire floods and debris flows.[8]

The U.S. Forest Service had exhausted its budget for firefighting before the Rim Fire even started. Cuts in most federal government programs triggered by "sequestration," automatic deficit-reduction measures that resulted from the 2011 debt-ceiling standoff between the Republican Congress and President Barack Obama, had whittled 5 percent from the Forest Service's $2 billion budget for dealing with wildfires.[9] Even without the sequester, federal firefighters were certain to run out of funding, but agency officials reported that the cuts meant they had 500 fewer firefighters and 50 fewer fire trucks than the year before.[10]

JUST BECAUSE THE FIREFIGHTERS were out of money didn't mean they wouldn't show up. They just had to get the funds from somewhere else.

"Fire borrowing" — the transfer of funds from other Forest Service and Department of the Interior programs to pay for firefighting — happened in all but four years between 2000 and 2013. In seven of those years, both the Department of the Interior and the Forest Service, part of the Department of Agriculture, ran out of money to fight wildfires. Part of that was due to skyrocketing costs. The USDA reported that fighting fires accounted for 13 percent of the Forest Service's budget in the 1990s, but consumed more than half of it in 2015.[11]

Caps on discretionary spending passed by Congress further hobbled firefighting budgets. In 2014 the U.S. Forest Service and Department of the Interior were projected to spend more than $470 million more fighting wildfires than the $1.4 billion Congress budgeted for the task.

"The forecast released today demonstrates the difficult budget position the Forest Service and Interior face in our efforts to fight catastrophic wildfire," Robert Bonnie, the undersecretary of agriculture for natural resources and environment, said on May 1, 2014. "While our agencies will spend the necessary resources to protect people, homes, and our forests, the high levels of wildfire this report predicts would force us to borrow funds from forest restoration, recreation, and other areas."[12]

For years critics have noted that the annual reallocation of funds is turning the Forest Service into the "Fire Service." A report in August 2014 from Tom Vilsack, the secretary of agriculture, cited a 110 percent increase in the number of people working on wildfires in the Forest Service since 1998, while the number of people managing forests had declined by 35 percent. "Fire borrowing" had led to a 95 percent reduction in the budget for maintaining infrastructure, including making critical bridge repairs, fixing health and safety problems in buildings and campsites, and maintaining water supplies — a $5.5 billion backlog of projects. Vegetation and watershed management suffered a 22 percent reduction; wildlife and fisheries habitat management a 17 percent cut; and support for recreation, heritage, and wilderness activities a 13 percent cut.[13]

Although Congress pays back most of the borrowed funds, Forest Service chief Tom Tidwell argued that this approach hamstrings programs to prevent and prepare for future wildfires, such as prescribed burning and thinning of forests,[14] while a letter from the Senate Committee on Energy and Natural Resources to Obama administration officials stated that "this approach to paying for firefighting is nonsensical and increases wildland fire costs."[15]

Political battles regularly leave firefighters under-resourced. In 2009 Congress passed the FLAME Act, which created a reserve fund to pay for additional firefighting in years with more blazes than average. Congress put $413 million in the reserve, but cuts brought it down to $290 million during the debt-ceiling standoff with the president in 2011.

While the FLAME Act fund was intended as a reserve to help cover the costs of expensive fire years, it was included in the overall wildfire budget based on the average costs of the previous 10 years. With the costs of wildfires rapidly escalating, the fund was quickly being depleted.[16]

FUNDING CHALLENGES WILL GET FAR WORSE as the costs of wildfires continue to explode.

A month after the sequester took effect in 2013, Forest Service scientists gave a grim prediction of what the "new normal" for wildfire in America will look like. At a forum in Denver considering how the nation should adapt to climate change, they released a paper projecting that by 2050 the area burned each year by wildfires will at least double, to about 20 million acres. The assessment, based on 25 years of climate science, predicts that some regions, including western Colorado, will see a five-fold increase in wildfires.[17]

Tree-killing insect attacks, expected to see similar growth, are already increasing, according to the report, killing high-elevation trees that were once protected by the cold. Runoff from fire scars will increasingly contaminate watersheds serving the booming population of the West. Forests' capacity to cleanse water will diminish, as will their ability to sequester carbon. Forests currently suck in about 13 percent of U.S. carbon emissions.

Scientists urging lawmakers to restore forests and reduce dangerous fuel loads instead found their budgets to prevent and prepare for wildfires slashed, in order to fund firefighting in forests that were already burning.

"This is not merely a concern for future generations; it will hurt us right now, this year," Vilsack wrote when releasing the 2014 report. "The Forest Service will soon run out of money and will be forced to transfer hundreds of millions of dollars from other programs in order to put out the fires . . . The current system is untenable, dangerous, and simply irresponsible."

AS COLORADO SPRINGS CONTINUED CLEANING UP from the two most destructive fires in state history, the political squabbling and fight for funding were just two more drivers of the new world of megafire. "I think we're looking at the tip of the iceberg," Steve Riker, the city's incident commander at the Waldo Canyon Fire, told me. "We don't see 99 percent of it. It is going to get worse.

"We keep pushing farther and farther into areas that we've never been into before. We're not allowing our forests to burn off like they should every hundred or 200 years. And the farther we get away from these events, our memories do get clouded. Are we going to get bigger, and worse [wildfires] and more deaths? Absolutely. I don't see any way around that."

That the city faced two once-in-a-century fires not even a year apart is a reflection of one aspect of the new normal: megafires aren't just blazes that are bigger or hotter or more deadly and destructive; they're also more frequent fires, out-of-season fires, and the endless fire season.

"It's the new norm," Jim Schanel said after he finished fighting the latest Colorado Springs conflagration in Black Forest. "Whether you subscribe to global warming, the population of the northern hemisphere increasing infringement on wildland areas, [or] unhealthy forests, there's a very complex algorithm to this. This fire behavior is not going to go away."

Within 10 days of the Black Forest Fire's containment, every aspect of that algorithm — the warming and drying climate, the overgrown and unhealthy forests, the homes filling flammable landscapes, the politics, and the economics — would come to bear on a small Arizona city.

PART VI

BACKFIRE

26

FRONTIER DAYS

Prescott, Arizona — July 1, 2013

PHOENIX MAY BE NAMED for the bird resurrected from its burnt nest, but Prescott is the Arizona city that was reborn in fire. That story is carved in stone.

The Yavapai County Timeline, etched into a concrete walkway in Prescott's Courthouse Plaza, starts with an explosion painted red, orange, and black, below the words "In the Beginning." A scarlet line jets from the blast like a highway stripe, at first in dashes that mark the sporadic history of white explorers in the area, and then unbroken, as if moving into a no-passing zone.

The stripe runs toward the courthouse until confronted by a statue of Buckey O'Neill atop his rearing horse. In Prescott, O'Neill was a newspaperman and sheriff. In Tombstone, he was a friend of the Earp brothers. And in Cuba, he was a captain with Teddy Roosevelt's Rough Riders, with whom he died on San Juan Hill on July 1, 1898 — 115 years to the day before I stood in front of his statue. He was shot through the mouth by a sniper after preaching that officers shouldn't cower from the Spanish bullets, and then rolling a cigarette amid the gunfire.[1]

The concrete timeline hairpins away from O'Neill's charge as if it is retreating from the cowboy in its path. Fires in the area sometimes seem to do the same thing — but not all of them.

In 1863 Joseph R. Walker established the first mining district in what would become Yavapai County. The gold rush was the "city's birth certificate," and the next year, at the encouragement of Congress and Abraham Lincoln, who wanted to secure its mineral riches for the Union during the Civil War, Prescott became a town.

The timeline doesn't turn into a yellow brick road, but instead remains a fiery red. Even more than gold, flames defined the city.

The timeline notes the Battle Fire in 1972 and the Castle Fire seven years later as the largest wildfires in the county's history, but Prescott's tragic, flaming legacy began a century before. That's written not only in the timeline but in every building I could see from it.

Brick and stone facades in the Romanesque, Second Renaissance Revival, Beaux Arts, Art Deco, and Chicago styles face the plaza. The square includes as distinctive a panorama of architecture as can be found anywhere in the West. When the town first rose amid the arid ponderosa pine forests, however, it was comprised almost entirely of wood-frame buildings. Settlers from the Midwest and East created a city in the style of those moister regions of the United States. Even the square, with its Neoclassical Revival–style courthouse surrounded by a park, reflects an eastern sensibility.

From the timeline, I looked up at Whiskey Row's famed lineup of taverns facing the courthouse. A longhorn skull nearly five feet wide spanned the doorway leading into Matt's Saloon. Its eyes glowed red, like embers. Beside it, the electric handlebar mustache of Hooligan's Pub matched the size and shape of the steer's horns. A Victorian lady outlined in neon led to the Jersey Lilly Saloon, its grand stone balcony overhanging the Palace, Arizona's oldest frontier saloon, a sprawling restaurant where tourists gawk at the bullet holes in the tin ceiling.

In 1879 the *Weekly Arizona Miner* newspaper saw a plague of fire coming to the city as clearly as I could see those barroom signs. "At least four deep wells should be made on our public plaza which might be the means of saving our town should a fire break out in the wooden buildings on Montezuma Street," the newspaper pleaded. "We can't afford a fire just yet."[2]

It was natural that the newspaper would take on "that all destroying fiend of fire." A blaze in the 1870s had damaged much of the publication's equipment. Another had burned its printing office, several stores, and the Palace. The newspaper's editor, Samuel N. Holmes, burned to death in his room at the Sherman House hotel in 1884. The paper's headlines called that blaze "The Severest Visitation of the Fire Fiend the City of Prescott Has Ever Experienced."[3] A few months later another headline reported a "Bath House, Laundries, Opium Dens and Chinese Joss House Swept Out of Existence" in "another disastrous fire."

On July 4, 1883, a fire destroyed many of the buildings around Courthouse Plaza, including much of Whiskey Row, despite the "fire wells" on all four corners of the square. Five years later, as a white building and a jagged bloom of flame carved into the timeline note, another fire destroyed much of downtown east of the plaza.[4] That same year, according to the next engraving, Prescott founded what is now the world's oldest rodeo. A list of the presidents of the rodeo are carved into the sidewalk, concluding with J. C. Trujillo, who made enough money riding broncos to purchase a ranch in the county and has been running Prescott's Frontier Days for decades. The rodeo had its struggles, and if it weren't for some of its supporters taking second mortgages on their homes to keep it running, it would have closed in the 1980s.

Prescott also has Arizona's oldest fire department, but its founding and fire chiefs aren't noted in the walkway.

The rodeo and fires might have been easy to confuse — often the first alarm for blazes was a volley of revolver shots. Other times, fires set off ammunition in burning buildings.

In 1889 arsonists started "two successful, and two unsuccessful" fires to break prisoners out of the Prescott jail, the *Miner* reported. One suspected arson burned down the plaza stables, killing 11 horses, on the same night that D. Levy & Co.'s barn was ignited with rags and coal oil. Later that year, in "A Notice to Fire Bugs," the paper announced the hiring of six night watchmen to stop the wave of arsons.

By mid-July 1900 residents watering their gardens had drained many city wells. The fire wells were out of use and covered. On July 14 the mayor prohibited the use of city water for irrigation, threatening to cut the supply to residents who didn't comply. Prescott was without water that day anyway, as mechanics repairing the system had disconnected the engine that ran its pumps.

It was Saturday night, and the saloons of Whiskey Row were overflowing with miners, ranchers, and cowboys. Roulette wheels spun, faro dealers called out their games, and piano players pounded out ragtime.

The most popular, if disputed, history holds that a miner stuck his pick candle into the wall of his room and forgot to blow it out before heading to Whiskey Row for a drink. Nobody disputes that the blaze on the "Night of Terror" raced easily through wooden bunkrooms and on to Whiskey Row. "The firemen are perfectly powerless," the *Arizona Weekly Journal-Miner* reported, "as there is no water."[5]

Harry Brisley, the owner of a pharmacy in the Burke Hotel, recalled "a scene worthy of a Western movie thriller," as the city engineer careened around the corner of Gurley and Cortez Streets on his way to the pumping station in a wagon pulled by his old white horse. "The slowly swelling hose proved the pumps were in motion," he wrote.

But it was, literally, too little and too late.

"The city faced the severest hours of its thirty-eight years' history, when from the nozzle came a dirty stream falling weakly to the street a few feet from the source," Brisley wrote. "By this time, fire had entered the . . . Scopel hotel."

Brisley reported (and others disputed) that firefighters threw dynamite into the blazing Scopel to try to collapse the building, but the masonry walls stood. Embers from the burning building landed on the wooden siding and shingles on the back side of Whiskey Row.

Firefighters dynamited other buildings, including the Burke Hotel at the end of Whiskey Row, in an attempt to create firebreaks. They had stopped the spread of at least one previous fire using this technique, but this time the flames easily leapt the gaps. Soon every building on the Row was burning. The entire block between Montezuma and Granite Streets burned to the ground. Gunshots rang out through the night as the flames set off bullets in stores and hotel rooms. Bucket brigades hardly slowed the flames; residents and shopkeepers saved whatever valuables they could. The huge, intricately carved backbar in the Palace, which had traveled to Prescott from San Francisco by boat and covered wagon, now rode on the back of the bar's patrons, who carried it across the street to Courthouse Plaza to save it from the flames.[6]

In less than five hours "the Great Fire" destroyed five blocks of the city, leveling 61 of its wooden buildings and doing nearly $1.5 million in damage. Only two of the buildings on Courthouse Plaza survived.

The day after the blaze, four bars opened in pinewood shacks in the plaza, and more came in the following weeks. The board of supervisors assigned each business that burned a site on the grassy square as close as possible to its ruins. Pianos, roulette wheels, and card dealers sang out from the park. And they had something to celebrate: nobody had died in the Great Fire of 1900.

Nearly every commercial building was reconstructed by 1901. But while wood shacks filled the plaza during the reconstruction, the buildings that rose from the ash were of stone and brick — architecture as fire resistant as it was grand.

But even that didn't stop the fires on Whiskey Row. Just a year before I visited, patrons sitting at the bar in the Bird Cage Saloon saw flames above the backbar. Hundreds of city residents, some of them weeping, spread across the courthouse lawn to watch flames leap behind the six-foot-tall Bird Cage sign.

A splash of fiery color fills an entire square of the timeline walkway to mark the Great Fire. Fires that transformed more than just the city are marked farther on. A mushroom cloud notes a Nevada nuclear test that shook Prescott, the nation, and the world. A few steps along, an air tanker carved in the concrete commemorates the first use of planes to fight wildfires.

As I stepped from Prescott's past into its present on Monday, July 1, 2013, the first day of the weeklong Frontier Days celebrations, both the rodeo and Whiskey Row were as healthy as ever. An elderly couple in matching western shirts sewn from American flags joined me on the plaza. Below the neon Harley-Davidson sign gray-bearded bikers gathered in black leather. Some downtown visitors displayed arms sleeved with tattoos, while a few others donned historic western garb. "Everybody's Hometown" lived up to its nickname.

In the coming days other unusual costumes filled the square. More than 100 bagpipers and drummers in kilts — the largest gathering of its kind since the terrorist attacks of 9/11 — marched along the timeline in front of the courthouse. Firefighters from around the world filled the Firehouse Kitchen and mingled around Old Firehouse Plaza, "Prescott's Hot Spot for Shopping and Dining."

But, as opposed to the rodeo fans and tourists, the firefighters and pipers in Prescott were anything but celebratory. On Jersey Lilly's balcony, some tourists in town for the Fourth of July rodeo cheered out a toast. "To the firemen!" they shouted.

I was confident that none of Prescott's firefighters were on the plaza to hear the tribute.

27

DEFUSING THE TIME BOMB

Prescott, Arizona — 1990

IN 1990, WHEN DARRELL WILLIS was Prescott's deputy fire chief, the former professional fast-pitch softball player kept fit by jogging through the city with Ed Hollingshead, who was in charge of dealing with wildfires in the Prescott National Forest. The patchwork of overgrown ponderosa pinelands and oily chaparral that surrounded the city combined to created a landscape as flammable as any in the country. The two firefighters would run through different subdivisions, bantering about the hazard.

"This house isn't going to make it."

"We can't deploy a crew here."

"We've got a problem here," Willis told his friend. "I don't think either one of us, whether it's the forest [service] or the fire department, is going to solve it."

That year a fire threatened the city a few weeks before the Dude Fire killed six firefighters near Payson, 100 miles to the east.

"That just sparked an interest," Willis told me.

Willis and Hollingshead founded the Prescott Area Wildland Urban Interface Commission, giving it the flaming logo "Living on the Edge." In 1991 they started holding evacuation and firefighting drills in Prescott's forested developments, but few residents took the threat seriously.

"We struggled for 10 years to get the community aware," Willis said. "I'd been telling the city these things, and they said, 'We don't believe you. We don't think it's a problem.'"

By 2000, when the National Fire Plan set funds aside to help communities prepare for wildfires, Willis was the fire chief in Prescott. He went to the city council to promote a wildland fire code requiring that homes that might confront a wildfire be built with flame-resistant materials and have defensible space cleared around them.

"The developers and the shake-shingle roof people all came out of the woodwork," Willis recalled. "They said, 'No, you're going to shut development down by increasing the cost.' So we have this battle . . . and in 2002 we have the Indian Fire."

That blaze nearly burned into the city.

"I went back in with the interface code, and I said, 'Is this proof enough?'" Willis said.

The city council adopted it, but angry developers later compelled the council to rescind its decision.

Willis brought in a consultant to analyze the wildfire threat to Prescott, and the city put together a committee of developers, contractors, firefighters, and residents to deal with it. By 2001 they had a community-wide vegetation plan. A grant from the state paid to clear some defensible space around private property.

Other grants funded the fire department's first fuels crew — a team of seasonal workers to cut chaparral and ponderosa pines from homes and developments at the edge of the city. By 2003 they had cleared defensible space around nearly 1,700 homes and removed 10,000 beetle-killed ponderosa pine trees. Willis couldn't get all the building codes he wanted, but he got some implemented for new construction in the most hazardous areas. In 2002 the city with the oldest fire department in Arizona finally got its Wildland Urban Interface Code, also the first in the state.[1]

But the city was hardly keeping pace with the hazard.

A risk-assessment company from California rated Prescott one of the West's 10 most likely communities to be burned over in a wildfire. Housing was sprawling fast. Even with a city fuels crew to do the work, residents were resistant to removing trees and brush around their homes.

Willis retired as fire chief in 2007, but came back in 2010 as chief of the city's Wildland Division, which he had created. Members of the division studied the threat to thousands of homes with a "Red Zone" computer program, inspected nearly 400 homesites a year for code compliance, and reviewed scores of building plans. At annual town hall meetings and in elementary school programs, they explained how defensible space would save homes.

But the city's greatest advancement in protecting itself from forest fires arrived in the form of a single man.

IN 2003 WILLIS HIRED a former hotshot to lead the fuels crew.

Eric Marsh was born and raised in North Carolina's Blue Ridge Mountains but fell in love with both wildland firefighting and Arizona when he joined a U.S. Forest Service crew during his summer break from studying biology at Appalachian State University. After graduating he moved to Arizona to fight fires full-time.[2]

Quiet, confident, and direct, Marsh had worked as a hotshot with a crew out of the Tonto National Forest before coming to Prescott. His affinity for the woods and penchant for pushing his body were the perfect combination for a career in wildland firefighting.

Marsh was an avid mountain biker who competed in 24-hour endurance rides, an equestrian with a beloved mount named Shorty, and a motorcyclist with a patch on his leather vest that read "American by Birth, Southern by the Grace of God!" A ski patroller, fisherman, and climber, he proposed to Amanda, his third wife, on an ice-climbing trip to Ouray, Colorado.[3] Sober since 2000,[4] he carried his drug of choice — a plastic bottle of Bisbee coffee along with a camp stove — with his fire gear so he could brew up in the woods or a parking lot. When he wasn't working or playing outside, he was cutting stone or welding steel in his effort to build a life in Arizona. Despite his taciturn demeanor, Marsh became a fixture in Prescott and eventually helped his parents, John and Jane, move from North Carolina to Arizona so they could be closer to their only son.

Shortly after his arrival in Prescott, Marsh and his second wife started the Arizona Wildfire Academy in the living room of their mobile home. Ten years later it was one of the largest and most respected training facilities of its kind in the country, drawing more than 700 attendees from across the nation. By the time the academy celebrated its 10-year anniversary in 2012, the National Fire Protection Association honored the Prescott Fire Department's wildland fuels management program as the "Gold Standard" in the nation.[5]

In 2004, just a year after Marsh joined the fire department, the fuels crew completed training as a Type 2–Initial Attack wildland firefighting hand crew, giving the city 20 dedicated forest firefighters during the height of the wildfire season. But Marsh and Willis had something more in mind. Their audacious plan to deal with Prescott's increasing vulnerability to wildfire was to turn the crew that most residents saw as a group of glorified landscapers into an elite Type 1 team of hotshots.

Nearly all of the more than 100 hotshot crews scattered around the United States were run by the U.S. Forest Service or the Department of the Interior. None had ever been part of a municipal fire department. But with increasing demand for hotshot crews to fight fires throughout the West, the city would get the protection of an elite team of wildland firefighters that, at least on the surface, was nearly cost neutral. The federal government's payments for the crew to fight fires outside their jurisdiction would pay for most of their work around Prescott.

Marsh dedicated himself to building his crew, pushing its members to improve their fitness and skills. Early in his time in Prescott, he got a reputation for being hard on people he disagreed with, but Willis urged him so much to be kinder that niceness became one of the guiding principles for the crew and its superintendent. Marsh, who had no children of his own, would come to be called "Papa" and would refer to his firefighters as his "kids." He gathered the crew for barbecues and drove them to doctor's appointments, but discipline was strict in his firefighting family. Even when covered with char, he made them keep their clothes and hair neat, with shirts buttoned up and tucked in. They addressed supervisors and homeowners with the same southern manners he was raised with.

"The first couple years [were] really rough because we hired a bunch of — and I was one of them — kids that really had no clue about life, much less fire," recalled Pat McCarty, who started with the crew in 2005. "Those guys just wanted to have a cool summer job and make some extra money and . . . go around saying, 'Hey, I'm a firefighter.'"

Others who saw all the elements of a disaster growing around the city joined in to help. Marty Cole, a Prescott firefighter with years of wildland experience, stepped off a fire truck to be a leader of the crew. Tim McElwee taught them techniques he'd learned fighting wildfires in California.

Within a year or so, crew members who weren't serious about the job began to drop out. "Eric just had a natural ability to get people to be honest with themselves," McCarty said.

On McCarty's first training hike with the crew, normally a four-mile hoof up a mountainside, a crew member carrying a chainsaw fell behind. Marsh grew angry and stopped the line. "I need somebody else to carry this saw," he said.

"I'll do it," McCarty replied, eager to impress.

"Rookie, don't you put this down until we get done with this hike," Marsh ordered. Then he extended the hike into a grueling 10-mile loop. All he said to McCarty at the end was "Good job."

"That next day, the guy who he took the saw off just goes, 'You know, this isn't for me anymore,'" McCarty recalled.

Marsh's crew stood out. "All those federal crews, they got green buggies. We pulled up in white vans and pickup trucks, and everybody's like, 'Well who are these guys?'" McCarty said. "'They work for a [city] fire department? Well, what do they know about wildfire?'

"Most of those crews were all older, grown men — full beards. We weren't allowed to have a beard."

The men that spilled out of the white trucks looked like they could still be in high school. But Marsh made sure they stood out on the fire line, too.

"Eric had a very competitive side to him," McCarty said. "Eric wanted to be the best hotshot crew in the nation."

He had something to prove to the Prescott Fire Department as well as to other wildland crews. "A lot of guys in our own department had a hard time getting used to these brush guys running around . . . with 'Prescott Fire Department' on the side of their trucks," McCarty said.

IN THE MIDDLE of the mountains of training and paperwork required to build his crew, Marsh drove to a view of Granite Mountain, pressed a piece of paper against the windshield of his truck, and traced the shape of the craggy peak. He turned the sketch into a logo and, in

2007, had it stenciled on the crew's trucks with a "T" to indicate the Granite Mountain Hotshots were in training.

He adopted a phrase in Latin as the crew's motto: *Esse Quam Videri* — "To Be, Rather Than to Seem." An essay he wrote explained it to his crew:

"We have a desire to continuously push ourselves physically and mentally, whether in training or on the job. We will exceed our perceived limits of what we can endure to tie in a piece of line, complete a hike or catch a spot . . .

"We always say Hi and act friendly. We remain humble and helpful at all times . . ."

A year later the crew was on a fire near Hells Canyon, in Oregon's Klamath National Forest, when Marsh got a call during the morning briefing with his men. He took a knife and peeled the "T" off the logo on his truck. "We're hotshots now," he said. "You guys earned this."

The crew had a quick round of high fives, then headed to the fire.

When city officials continued to make it difficult for the crew to meet its requirements, Marsh wrote another essay explaining his crew to the community.

"To our peers, the 110 other Interagency Hotshot Crews in the nation, we are an oddity. In a workforce dominated by Forest Service and other federal crews, we have managed to do the impossible; establish a fully certified (hotshot) program hosted by a municipal fire department . . . We look different. Not because our buggies are white instead of green, but because we smile a lot. We are different. We are positive people. We take a lot of pride in being friendly and working together, not just amongst ourselves, but with other crews, citizens, etc. . . .

"To our city coworkers, we are a bit of a mystery. Guys that work in the woods a lot . . .

"To our families and friends, we're crazy. Why do we want to be away from home so much, work such long hours, risk our lives, and sleep on the ground 100 nights a year? Simply, it's the most fulfilling thing any of us have ever done . . . We don't just call ourselves hotshots, we are hotshots in everything that we do."

Marsh was quiet about his pride in his crew, declining even to have them march in Prescott's parades.[6] But it showed on his truck, which would soon have custom Granite Mountain taillights. Five years after he scraped the trainee sticker off of it, most of the cars and trucks in Prescott would be decorated with the logo Marsh had designed. The circumstances, however, would be anything but celebratory.

IN THE MID-1990S CONRAD JACKSON was on the Prescott Hotshots, a Forest Service crew based outside the city. He studied science and education in college and was a substitute teacher in the off-season. When Prescott High School offered him a full-time job, his work obligations flipped. He taught chemistry during the school year, then worked on wildland fire engines in the summer. In 2000 he took a retired fire truck and used it to teach a firefighting class he created at the high school — another nod to the city's outsized wildfire hazard.

Shortly before 3 p.m. on May 15, 2002, Jackson was preparing for that evening's firefighting class when some of his students ran in to tell him about a forest fire that was pushing hard toward the city. They gathered on the high school's patio to watch a constant stream of air tankers swooping at the Indian Fire four miles to the south. Class was canceled when Jackson was called to join the fight. By the end of the day, five homes had burned. Incident commanders, however, had feared that the blaze could destroy 2,000. The preparations for a wildfire in the city had paid off.

"We had a written plan, and we drilled on it for 12 years," Darrell Willis told the *Prescott Daily Courier*.[7]

A "brush crush" organized by the Prescott Area Wildland Urban Interface Commission two years earlier had left 300 acres between Cathedral Pines and Quartz Mountain thinned enough that the crown fire dropped to the ground, providing hand crews a place to make a stand. The Prescott Fire Center, with aircraft and retardant, was just five minutes from the blaze, and the Prescott Hotshots based there happened to be home. With the fire season just starting, there was little competition for resources. Four heavy air tankers and four hotshot crews were on the blaze within an hour of it igniting.[8]

Still, the losses would likely have been even lower had the community been more cooperative. A prescribed burn just south of Cathedral Pines three years earlier had been shut down due to complaints about smoke from residents. Firefighters managed to burn only 30 of the 120 acres they planned to treat.[9]

From some members of the public, there was open hostility toward firefighters who set blazes or cut down trees to reduce hazardous fuels. "When we had the prescribed burns, we actually saw letters to the editor saying we should go out and kill Forest Service employees," Jackson told me.

Firewise preparations saved some homes in the Indian Fire, but one burned after the woodpile beneath its deck ignited. Luck was a big part of the firefighters' success. The blaze started late in the day, so the winds died down and the weather cooled just as it was building momentum. Dozens of slurry drops from planes allowed firefighters to cut lines that corralled the blaze by midnight.

A HANDFUL OF STUDENTS FROM JACKSON'S FIREFIGHTING class joined Eric Marsh's fire crew.

Pat McCarty planned to be a pilot, but he joined the Granite Mountain Hotshots, then took a job on the structure firefighting side of the Prescott Fire Department. His younger brother, Daniel, who was known as "Booner" on the crew, followed in his footsteps.

After graduating from high school, Travis Turbyfill joined the U.S. Marine Corps, where he was a machine gunner and marksman. When he came home, he joined the hotshots. Ten years after taking the high school firefighting class, he was a squad boss with the crew, now married with two blond daughters.[10] Andrew Ashcraft, another graduate of Jackson's class, had the names of his four kids tattooed on his chest. A Mormon, he had a wristband reminding him to "be better," but he was the same wisecracking spark plug he'd been in high school.[11]

When the city decided to require teachers to work year-round, Jackson had to choose between his job at the high school and his summers fighting fires. He took a full-time position with the Prescott Fire Department but kept teaching his fire science class. A few years later enrollment in the class dwindled, and Jackson gave the high school's fire truck to Yavapai College for its firefighting class. But he continued teaching firefighting to Boy Scout Explorers, and one of his students there, Brendan McDonough, would also end up on the Granite Mountain Hotshots.

Jackson wore a U.S. Forest Service belt buckle, sometimes with a golf shirt from the Prescott Fire Department — a reflection of his devotion to both wildland and structure firefighting. He kept his hair cropped close and his body honed with CrossFit classes, where he often ran into some of the hotshots.

The last time he saw any of them other than McDonough, he had popped into Station 7. The crew's plain, gray base camp had a workshop, gym, and tile floor spelling GMIHC, for Granite Mountain Interagency Hotshot Crew, inside. Marsh had laid the tile himself, and any rookie who stepped on one of the letters had to do push-ups as penance. Jackson had dropped in to see who on the crew needed their picture taken. It was a point of pride to have the crewmembers' photos hung alongside those of the structure firefighters in the department's offices.

Cumulus clouds were building in the sky, and the hotshots, knowing his love of weather, ribbed him.

"Hey, Jackson," one shouted to him. "Is it going to rain today?"

He knew they weren't asking for a weather report, but he never passed up a chance to teach.

"Well, what's the dew point?" he responded.

"The [relative humidity] is . . ." the firefighter started to respond.

"I don't give a fuck what the RH is," Jackson interrupted. "What's the dew point?"

The hotshots admitted they didn't know.

"So they got this little micro-lesson because they made the mistake of asking me 'Why do you give a fuck about the dew point?'" he told me. "That was my last interaction with the entire crew."

To the rest of the crew, Brendan McDonough was "Donut" — a shortened version of his last name. After taking Jackson's fire science class, he'd had a few wild years with plenty of drinking and doping. He spent a couple of nights in jail in December 2010 for his involvement in the theft of a radar detector and GPS from a car,[12] and three months later his girlfriend gave birth to a daughter. The next month he showed up at Station 7 to ask about a job. He'd always wanted to be a firefighter and had heard Granite Mountain was hiring. Some of the firefighters on duty who knew his reputation told him there weren't any openings, but Eric Marsh overheard the conversation and caught Donut before he left.

"Can you do an interview?" he asked. ". . . Right now?"

Donut was wearing a dirty white tank top and grungy work pants, but he sat down with the super. Marsh asked a question he posed in every interview. "When was the last time you told a lie?"

Donut told him about smoking pot and his record and his daughter. He had all the appropriate training and certifications, but he had never fought a wildfire. Marsh hired him.

From that moment forward, Donut revered Marsh as the man who gave him a second chance and drove him hard to make good on it, not just as a firefighter, but as a father and a member of the Granite Mountain family.

CLAIRE CALDWELL FIRST SAW HER HUSBAND, Robert, on Prescott's Whiskey Row. It was '80s night in a dive called Coyote Joes.

"They had . . . full-on neon short-shorts and sweatbands," Claire recalled of the two hotshots spinning on each other's shoulders on the dance floor. "They would run into the girls and not even care."

Some of her friends were put off, but Claire, a single mom with a love of the outdoors, laughed at the horseplay.

A few weeks later, back on Whiskey Row, a friend introduced her to Robert. She told him about backpacking 100 miles around the Grand Canyon. He told her about being a hotshot.

"He had been places where no man had ever walked," she told me. "I thought it was the most attractive thing in the world."

Robert took Claire's hand to find a quiet corner where they could keep talking. They went home together and exchanged phone numbers. When they arranged another date at Coyote Joes, Claire went with her girlfriends and Robert came with some hotshots, but they left together after about half an hour. From that point on, when Robert was in Prescott, he and Claire were inseparable.

He warned her that his job might keep her from seeing him, or even speaking with him, for weeks. "I don't choose your career, your path, your destiny," she told him. "But I'm in love with you, so I accept you 100 percent for who you are and what you do. I will be your girl."

A few weeks later, at midnight, Granite Mountain got its first out-of-town fire assignment of the season. "He put his boots on, and it was literally fire mind all of a sudden," she said. "Boom, he was gone."

They got engaged two months after they met.

ALWAYS COMFORTABLE LIVING ROUGH, Robert seemed to grow happier with each day he slept in the dirt and went without a bath. During spring break of his senior year in high school, he went to the Arizona Wildfire Academy that Eric Marsh had founded.

"He realized it was exactly the kind of thing he wanted to do," his friend Tom Holst told me. "The academy was all he could talk about."

Traveling to exotic and rugged landscapes, camping, working hard high in the mountains and deep in the woods — Robert had found his life's calling. After finishing his training, Robert kept his bag packed at the foot of his bed, waiting for the chance to fight a fire.

Robert's father, Dave, called Brian Misfeldt, who had a small company that worked on wildfires. Brian said that he would hire Robert if the teenager could walk three miles in 45 minutes wearing a 45-pound pack — the infamous pack test required of every wildland firefighter. When Brian saw the whippet-thin teenager get out of his truck at the Prescott High School track, he thought some firefighting friends were playing a joke on him.

"Here comes Waldo," he thought. "This isn't going to be good. He's got these skinny arms and skinny legs and . . . a big old grin on his face."

Brian hung a vest with weights on Robert. "Can you hold this up?" he asked.

Brian's wife, Renee, a Forest Service firefighter, makes sure he keeps precise records. He wrote down the time as Robert took off walking fast.

"You might want to slow down," he yelled to Robert. "You've got 12 laps to go."

"At the 12th lap he still had that big old grin on his face and one of the best times I've ever seen," Brian recalled.

Brian and Renee's house was like a fire camp, filled with fire packs and Pulaskis. It was soon a second home for Robert.

When Brian had started his business, he'd flown to the East Coast and bought an old, decrepit fire truck. Robert went to work on it as soon as Brian hired him.

"He was a fantastic mechanic," Brian said. "The kid could do anything with electronics. We always had problems, and he loved to fix them."

Brian and Renee each had reasons to prioritize safety with their workers. Renee had lost a good friend in the 1994 disaster on Storm King Mountain. That same year a fire had trapped Brian and two other firefighters in a steep bowl of chaparral and cheatgrass in the Mackenzie Fire, about 40 miles southwest of Kingman, Arizona.[13] "I turned and it was a wall of fire," Brian recalled, "from one end of the bowl to the other."

They crawled under their fire shelters in a boulder-strewn gully. "There was so much noise from the fire it was unbelievable. It was just a tremendous rumbling," he said. "And then it hit. It was on our back. I remember the pain getting really bad."

Brian managed to stay beneath his shelter, despite the burning and fear that made him want to run. "I had the boulder right there, so I was beating my head against it . . . trying to knock myself out. My lungs started burning. I didn't want to breathe anymore."

Brian's shelter delaminated. He could feel his pants burning off his legs. The flames consumed so much oxygen that he couldn't get a breath. His friends took a photo of him when he finally climbed out from under the shelter. "Because of the carbon in my lungs, my whole body had swelled up . . ." he said. "My eyes were slicked closed, my face was bubble round."

Brian worked with a veteran of the deadly Dude Fire who had drilled into him that, as painful as the flames might be, the only chance of surviving a burnover was to stay under the fire shelter.

"And I forced that into Robert's head," he said. "Stay there. Do not leave. Do not run no matter how bad it gets."

Brian was sad to see Robert move on, but proud to see him become a hotshot and happy to run into him at fires.

ZION, CLAIRE'S SON, took to his firefighting stepfather. While most of the hotshots did CrossFit on top of their punishing training hikes and runs, Claire told me that Robert and Jesse Steed, the captain of the crew, did "kidfit"—bench-pressing their children or doing pull-ups with them hanging on their backs.

The hotshots increasingly brought their wives, children, and girlfriends to barbecues and parties. Claire hit it off with Amanda Marsh, Eric's wife, because they had both gone to Prescott College. "We should start the Granite Mountain Girls," Amanda told her. "We'll be there for each other."

When the crew was returning from a long time away, Claire and some of the other GMGs would shop for food and lingerie together to make sure their men got a warm homecoming. Sometimes they would gripe about the crew's erratic schedule and how hard it was to have them away for so long. "Welcome to a lifetime of missing Robert," Claire wrote in her journal.

When the hotshots traveled, they saw some wildfires fueled by overgrown or unhealthy forests, others exacerbated by climate-driven drought, and still others complicated by development. Sometimes political and economic factors added fuel to the fires.

But when they came back to Prescott, they seemed to land at the center of a bull's-eye, where all those factors came into play. Arizona was warming and drying faster than any state in the contiguous United States.[14] Most of the local woodlands were desperately overgrown.[15]

When Robert came home from faraway assignments, he would show off photos of sunsets and mountain vistas, but after nearby fires he would talk about overgrown forests and explosive fire behavior. Claire was also most troubled by the local fires, which she could sometimes smell or see.

He warned his wife about fires in chaparral. "They're the scariest because they burn so hot and so fast and there's so much of it, it's hard to clear," he told her.

The manzanita and catclaw were all but impenetrable and tore at firefighters' clothes and skin. Central Arizona's ponderosa pine forests, exponentially denser than they would have been naturally, were prone to explosive crown fires. A pair of photos hung on the wall of the hotshots' station. The first, from early in the twentieth century, showed a parklike forest outside Prescott, with trees dozens or hundreds of feet apart. A photo of the same site from a century later, after the forest had been seeded to help grow Arizona's timber industry and every fire there had been extinguished, showed an almost impenetrable wall of trees.

When Robert would drive with Claire on excursions, he would point to the hazards. "That's going to burn," he would say, nodding to the thick chaparral. "And that's going to burn," he would add, pointing to the pine forest beside it.

The surge of vegetation was matched only by the speed at which houses were springing up in it. And the people building those houses were often resistant to anyone telling them how to build, what trees to cut down, or that they should invest in their protection from wildfires.

"The hotshots had a sense of pride — they were the protectors of this area," Claire said. "A lot of people would look at them and be upset they were chopping down the trees. People in the city would say they were living off taxpayers' dollars. We would get stuff like that a lot."

In 2013 Robert was promoted to squad boss. "When he found out, he called me, he called my dad — he called like 15 people within the course of an hour," Tom Holst recalled.

Robert had found a job that was everything he'd ever dreamed of doing with his life, allowing him to spend as much time as he could in the woods, train like an elite athlete, travel to places few humans ever saw, and develop a skill set that was part lumberjack, part scientist, part soldier. Most hotshots see the job as a stepping-stone to something else — college, off-season adventures, a job with a city firefighting department. But Robert never wanted to be anything but a hotshot.

"Don't ever ask me to quit," he told Claire when they first met, and she promised she never would.

But when he and Claire started talking about having a baby, he said he'd start looking into getting a job with the fire department that would keep him in Prescott.

In the meantime, both Robert and Claire helped grow the hotshot crew.

When Claire met Robert, she was working at a Mexican restaurant down the street from Whiskey Row. A week later Grant McKee, Robert's cousin, started working at the restaurant, too. He had grown up in California, but he had often visited Prescott, where his grandparents owned a clothing store. He moved there when he was in 10th grade and lived with Robert and his parents. The wiry, confident teenager played football, boxed, wrestled, and ran a marathon. Robert encouraged him to join the hotshots. Grant applied in 2013 and got a spot when another crew member bailed out. He didn't plan on fighting fires for long. He'd completed his EMT training and was looking forward to a career in emergency medical care.[16]

Later, the restaurant hired a handsome young dishwasher named John Percin, who had moved to Prescott from the suburbs of Portland, Oregon. "You should work with my husband," Claire told him.

Percin saved up his money from the restaurant to attend the Arizona Wildfire Academy. Known as "Coach" to many of the other hotshots, Percin was devoted to his English Lab, Champ.[17]

McKee and Percin, like many in Prescott, had cleaned up their lives in the desert oasis and were known for their work on behalf of anti-drug programs. McKee spoke at D.A.R.E. presentations throughout the county,[18] while Percin remained close with the staff of Chapter 5, a local addiction treatment facility that would later found a scholarship in his name.[19]

BRIAN MISFELDT RAN INTO ROBERT in fire camp at the Doce Fire in June 2013 and saw how far the gangly teen had come.

"I'm totally in love," he told Misfeldt. "It's the best thing that's ever happened to me . . . Let me show you my little boy."

Robert was leading a crew of hotshots but still had the same goofy grin, now framed by a mustache that offset his balding head.

"Hey, you got any words of advice for me?" a teenager Misfeldt was training asked Robert.

"Yeah," he responded, nodding toward his mentor. "You just got to be really understanding and patient with these guys."

The firefighters laughed into the smoky night.

28

THE DOCE

Prescott, Arizona — June 18, 2013

THE DOCE FIRE IGNITED on June 18, 2013, just 12 days before the tragedy on Yarnell Hill.[1] It started in an area popular with target shooters, near where Eric Marsh had traced Granite Mountain for his crew's logo. By the end of the day, it had burned over the peak the hotshots had been named for.

"It was like a kick to the stomach," Brendan McDonough said. "Like getting rid of a mascot for a famous sports team. That was our name. That was who we were. That's how we were remembered. That was our mountain that we hiked. That was us . . . It didn't feel good to see it burn."

But there was more at stake than their namesake. Beyond the mountain, Donut said, "million-dollar homes . . . are about to get nuked."

The nation's only two very large air tankers (VLATs) painted nearly 12,000 gallons of retardant with each drop onto the flanks of the fire, helping crews hold the blaze back from the houses.[2] The Granite Mountain Hotshots weren't there and had a far less dramatic rescue mission. They focused on keeping the blaze from spreading beyond their mountain. While that meant most of it burned, they took consolation in saving a single tree. The second-largest alligator juniper in the country resides on Granite Mountain. They dug a line and burned out around the eight-foot-wide trunk. When they returned to find it had survived the fire, they took celebratory photos of themselves stacked in human pyramids and hanging from its branches.[3]

The photos showed a tight crew. There was the squad boss and sawyer Travis Carter — quiet, willing to take on any task, and nicknamed "Honey Badger" for his relentlessness digging and sawing.[4] Anthony Rose had come to the crew from the Crown King Fire Department,[5] and Joe Thurston from the Groom Creek wildland crew.[6] In Prescott both worked as swampers — clearing away timber felled with chainsaws — and often commuted to work together. Travis Turbyfill was Paul Bunyan strong, but brought a copy of *Goodnight Moon* with him to fires so he could read it to his daughters over the phone from the fire camp.[7] In 2012 he won the crew's "Big 7," awarded to the member who had the biggest heart.

"Turby" and the crew's captain, Jesse Steed, were thrilled to have Billy Warneke, another marine, join the crew at the beginning of the 2013 season. He was expecting his first child in December.[8] They were equally excited that they'd have another father working with them. Sean Misner was also in his first season with the crew and was expecting his first child in September. His great-grandfather, grandfather, two uncles, and a cousin were all firefighters in California.[9] Dustin Deford was one of five firefighting brothers from a Montana family with 10 children.[10]

Kevin Woyjeck, the son of a captain in the Los Angeles County Fire Department, and the nephew of two other firefighters, grew up knowing he was going to be a firefighter, too. He was a Los Angeles County Fire Department Explorer and an EMT with an ambulance company after high school. He broke up his training with country line dancing on the weekends.[11]

Garret Zuppiger's full, red beard fit with his natural affinity for fighting fires and working in the woods, and made other crew members happy that Marsh had eased up on his restrictions on facial hair.[12] Scott Norris worked the off-season at a gun store, wrote witty and poetic emails describing his travels around the world, and loved studying weather.[13]

Chris MacKenzie had helicopter certification and calculated all the crew's loads for their chopper rides.[14] "He had this ridiculous, big, dumb calculator wristwatch straight out of the 1980s that he always wore," former Granite Mountain crew member Doug Harwood recalled in a tribute. "We loved to tease him about that watch. He, of course, thought it rocked."[15]

MacKenzie was sincerely interested in friends and strangers. "Chris had a knack for always making every conversation all about you," Pat McCarty wrote.[16]

Clayton Whitted saw his job as squad boss not unlike his career as a youth minister. He was a man of God first, a hotshot second.[17] "He could connect with anyone," McCarty wrote of him. "His desire to make people happy or smile extended to everyone on the crew. Whether it was letting us drown him, shave his head, or eating something gross — he'd do it for the other guys. His Bible was covered in the names of everyone he prayed for — that included every crew member he ever worked with."[18]

Eric Marsh, the leader of the hotshots, was just getting back in the action. He had been injured in a mountain biking accident and had been on "light duty" for the first couple months of the season. He was assigned as a division supervisor at the Doce, so he still wasn't really back with his crew.

MARSH KNEW THE CREW as a whole was a bit hobbled as well.

Most of the Wildland Division's funding came from grants and in reimbursements from the U.S. government for the Granite Mountain Hotshots' work outside the Prescott department's jurisdiction. When the hotshots worked on federal or state land, the city was paid $39.50 an hour per firefighter. Such an assignment could bring up to $200,000 into the city's coffers. With the crew responding to more than 150 out-of-town incidents in the previous five years, that money had added up. The city had planned on the crew being reimbursed up to $1.5 million during each of the three previous years. In 2011 the hotshots' operating costs had required only $6,975 from the general fund. The following year they had run a deficit of $68,340, but payments for fighting fires for the feds still covered more than 95 percent of the crew's costs.[19]

In 2013 increasing wildfire activity around the nation was bringing in more federal and state money, moving the crew from the red toward the black. That year the city expected the hotshots to bring in $1.3 million, but their actual reimbursements were headed toward $1.6 million. The city had a nearly $240,000 surplus over its investment in the crew, not counting its costs for benefits, capital expenses such as trucks, or the matching funds it paid for grants to thin hazardous vegetation around the city.[20]

Part of that surplus, however, was from cost cutting by the city. In 2012 Prescott had eliminated two of the full-time positions on the crew. Those cuts, Marsh knew, presented the Granite Mountain Hotshots with an existential crisis. Among the standards for Interagency Hotshot crews was one requiring that at least seven of the firefighters be in "permanent/career" positions. After the staffing cuts, the crew had difficulty meeting that requirement.

On a checklist provided to the Southwest Coordination Center in Albuquerque that spring, the Prescott Fire Department indicated that it did have the seven required permanent firefighters on the hotshot crew, listing Christopher MacKenzie as filling the last spot.[21] MacKenzie's personnel file back in Prescott, however, indicated that he was a "temporary and seasonal" employee. And while some seasonal employees could count toward permanent/career, such employees generally qualified for most of the benefits afforded full-time firefighters. MacKenzie, however, had signed a "temporary employment acknowledgment" making him ineligible for many benefits, including health insurance with the city, as well as paid sick leave, holidays, and vacations.[22] Earlier in 2013 the fire department, with city approval, had reclassified MacKenzie and Andrew Ashcraft as "full-time temporary employees," which kept the crew in compliance with the federal rules. Around the same time, Ashcraft was promoted to "lead saw," a position that had always qualified for full-time benefits but, due to city budget cuts that had frozen positions on the crew, wasn't supposed to this time.[23] In all, 14 members of the crew were temporary seasonal employees who made between $12 and $15 an hour and had none of the benefits afforded full-time employees. The hotshots were sometimes short of the required number of "senior" firefighters qualified to fill command positions, but MacKenzie's records showed that just five days before he and the rest of the crew headed to the Yarnell Hill Fire, he had completed the training and been recommended for certification as a Type 1 firefighter and Type 5 incident commander, which would have qualified him for that status.[24]

It's not uncommon for a hotshot crew to fail to meet all of the staffing and training requirements, and they are occasionally knocked down from Type 1 to Type 2 status while they deal with them. But to Marsh, these were no small shortcomings. In his own job review that May, he noted, "It is challenging to run a nationally recognized program with minimum standards and requirements that I am unable to meet . . . I believe things are starting to change; however, I still have some big questions that need answering about staffing."[25]

Wildland chief Darrell Willis responded to Marsh's concerns about staffing and certifications: "We have all spent a lot of time and energy trying to fill the positions. It's now time to let the system work, realize that we have done our best, and make the best of the situation."[26]

There were other challenges in keeping key positions filled, Willis noted, citing a "major disruption in staffing . . . just a few days prior to the season[al] firefighters starting."

Two years earlier, tensions between Marsh and his captain had led the captain to resign. The year before that, the crew had had an "extraordinary situation with one of our supervisors that ended up with a resignation," Marsh noted in his year-end evaluation.[27]

The federal certification checklist showing that Granite Mountain had the appropriate number of permanent employees and senior firefighters required a signature from the crew's superintendent. In 2013, however, Marsh didn't sign it. Jesse Steed, who had stepped into the superintendent role while Marsh was on light duty, signed it instead.

By the middle of June, Marsh was back on the roster and eager to work a fire. While he and his crew were on the Doce Fire, the National Interagency Fire Center sent a special advisory regarding fires in Arizona and New Mexico. Years of drought had resulted in fuels that could blow up quickly and burn through the night. "Firefighters should acknowledge that fire growth and fire behavior they encounter this year may exceed anything they have experienced before," the advisory warned.

AFTER THEIR TIME ON THE DOCE FIRE, the Granite Mountain crew was just a few days away from mandatory time off, but until then, they were on call. Wade Parker had tickets to see his favorite Christian rock band, Casting Crowns, in Prescott Valley, but texted his fiancée, Alicia Owens, that he couldn't go. "We've got to stay in this dumb-ass station," he wrote. "We can't go eat or do anything because there's lightning in the area."

He cheered up when Andrew Ashcraft's wife, Juliann, brought cookies by the station, which she often did when the hotshots were there. It had been Parker's turn to pick the flavor. He'd asked for chocolate chip–bacon.

Late in the afternoon of June 28 the Prescott National Forest called in the crew for help with a fire on Spruce Mountain, just up the road. "Hey, babe," Parker texted to Owens. "We popped a little fire out by the last one. Sorry I wasn't able to let you know. I love you."

29

WHERE THE DESERT BREEZE
MEETS THE MOUNTAIN AIR

THE LIGHTNING PARKER COMPLAINED OF marked the annual monsoon's arrival in Arizona. Each summer moist air from the Gulf of Mexico pushes in from the south, displacing the dry winds that come from the west in the spring. Thunderstorms form from the wet air and desert heat. The early ones bring lightning and wind, but rarely rain.

On Friday, June 28, the first of the monsoon storms formed on the Mogollon Rim, the rugged escarpment that runs diagonally across central Arizona to southwestern New Mexico to form the southern edge of the Colorado Plateau. From the rim, the storm moved west, passing over Prescott and peppering the mountains with dry lightning. At least seven of those strikes started fires, bringing the total number of wildfires in the state to 37.[1]

At 5:36 that evening lightning struck the top of what would become known as Yarnell Hill, a peak in the 6,000-foot-high Weaver Mountains southwest of Prescott. The rocky slopes' chaparral — manzanita, scrub oak, catclaw, and juniper — hadn't burned in nearly 50 years and had grown impenetrably dense. Extreme drought and 105-degree temperatures primed the thick, boulder-strewn scrub to burn. Citizens in the town below, which had received barely two inches of rain so far that year, had been warned that they faced extreme wildfire danger for more than a month.[2]

BARBARA KELSO, 94, had spent seven years on the Yarnell Fire District Board, but had retired six months earlier. She was out for dinner with family in Peeples Valley, north of Yarnell, when she saw lightning strike the mountain and smoke start to rise.[3] She immediately called 911.

Lois Farrell was sitting on her porch looking out on the Weaver range when she saw the lightning strike. Her husband, Truman, a 73-year-old U.S. Air Force veteran, had been the chief of Yarnell's volunteer fire department until two years earlier. She pointed out the smoke to him, but neither of them were worried by it.

Four minutes after the lightning strike, the Congress Fire District's volunteer fire department just west of the Weaver Mountains reported the fire to the Arizona Interagency Dispatch Center. A plane from the Doce Fire, still smoking near Prescott, detoured over Yarnell Hill and reported that the fire there was less than an acre in size, smoldering in a boulder field.[4]

A spot weather forecast that night predicted temperatures up to 104 degrees on Saturday, with low humidity and light westerly winds. If thunderstorms developed, they would likely bring wind but not rain.[5]

Yarnell, a former gold-mining town, barely hangs on to U.S. 89, which climbs through the center of town and then plunges into the Sonoran Desert nearly 2,000 feet below. When the fire started, residents could see the wispy smoke from the sleepy main drag, which contained a few antiques and collectible shops, a library, a senior center, a soon-to-close grocery store, and the Ranch House Restaurant. Northwest of the quiet town center, however, hundreds of homes hid on the dead-end roads weaving through dense chaparral and boulders.

"Where the Desert Breeze Meets the Mountain Air" is the town's motto, but some locals proudly describe Yarnell as "Gays, Grays, and Strays." Nearly 40 percent of the population is 65 or older, and the community was once popular with same-sex couples.

One resident I spoke with used different Gs to characterize the town. "It's all God and guns around here," he told me of a population that "just wants to be left alone."

In 2012 more than 100 people participated in the town's progressive Christmas dinner, and residents routinely banded together to provide meals to sick neighbors or help financially strapped residents pay their utility bills. The town's biggest attraction, the Shrine of St. Joseph of the Mountains — featuring stations of the cross set along a path climbing through a maze of granite knobs — was beloved by Yarnell's citizens, regardless of their religious persuasion.[6]

But the community of about 640 permanent residents and 100 summer homeowners was less cooperative in their preparations for wildfires.

Jim Flippen and his friend Don Mason also saw the lightning strike on Yarnell Hill on June 28, 2013, and the smoke that rose from it. Flippen had lived through the Oakland Hills firestorm in California in 1991, which had killed 25 people in a neighborhood thick with invasive grasses and imported eucalyptus. "It destroyed some of the nicest homes in San Francisco," he recalled.

When he moved to Yarnell, he kept a 25-foot perimeter cleared of vegetation around his house and had a metal roof. But some of his neighbors did nothing to prepare their properties for wildfires. "Our place would not have burned if our neighbor had done anything," he told me after he lost his home and Mason lost two houses and his blacksmith's shop in the Yarnell Hill Fire.

More than two-thirds of the land that burned in the fire was privately owned, and almost all of the rest was state property. Using satellite imagery, the Pacific Biodiversity Institute, in Winthrop, Washington, estimated that 89 percent of the homes in Yarnell had direct contact with the vegetation surrounding them. Only 63 percent of the buildings had any defensible space at all.[7]

Firefighters in central Arizona had warned about the potential for a disastrous blaze there since the 1970s. One former chief of the Yarnell Fire District had urged citizens to clean up their properties, but few had heeded the warnings. Instead, state and federal grants paid for prison crews to thin what they could. The year before the lightning strike, when the BLM offered to clear brush and trees around homes in Yarnell, only four people took advantage of the offer, and the Yarnell fire chief left a $15,000 grant for brush clearing unspent because, he said, he couldn't get enough volunteers to do the work.[8]

30

THE PERFECT FIRESTORM

RUSS SHUMATE, A FIRE MANAGER with the Arizona State Forestry Division, came on Friday night as the incident commander for the Yarnell Hill Fire, but chose to take no action on it until Saturday. "Not much of a threat," he told the state dispatch center. Volunteer firefighters from Peeples Valley, northeast of the fire, knew a trail that led to the fire and asked for the go-ahead to hike up and deal with it. But they would be working at night, in rugged terrain and heavy, volatile fuels. Shumate told them to stay off it.[1]

He instead requested two crews of inmates from the Lewis and Yuma state prisons, a wildland fire engine, and a light helicopter to attack the blaze on Saturday morning.[2] Some firefighters and residents would lament the delay in attacking the fire, which might have been easily snuffed on Friday, when it was small and the weather conditions weren't yet driving its growth.

By sunrise Saturday the fire was between one and eight acres in size and was the only lightning ignition from the previous day that was still burning. More than 40 firefighters from the prisons — inmates, crew bosses, and corrections officers — arrived early that morning, but incident command couldn't find a way for them to get to the blaze. It was around 11 a.m. when six members of the Lewis state prison crew, along with one BLM helitack crewman, choppered in and began putting out hot spots along a two-track jeep road on the eastern edge of the blaze.[3] The rest had to watch from Yarnell when the helicopter pilot determined that the landing was too dangerous to fly anyone else in.

Two single engine air tankers (SEATs) made several slurry drops around the fire. By early afternoon firefighters on the ground and in the air had contained it at six acres. It looked like the small crew had it beat, but steadily rising winds and heat awakened the sleepy blaze. Around 4:30 it got into an island of unburnt chaparral and slopped over the two-track road that was holding it. It was outrunning the firefighters trying to corral it and burned over their Gatorade, food, and other supplies. Incident command called for a large helicopter and a heavy air tanker to slow its eastward spread, but high winds kept the aircraft grounded.

The blaze followed the ridgeline northward, growing to 100 acres by 7:38 p.m.[4] During the night it spread to within a mile of the south end of Peeples Valley and 1.5 miles northwest of Yarnell.[5] The fire that nobody had thought much of the evening before was suddenly serious business.

That night incident command ordered fourteen Type 6 wildland fire engines, six water tenders, two bulldozers, three heavy air tankers, four more SEATs, six helicopters, and several crews of firefighters. The incident commander asked for a "short" Type 2 Incident Management Team, including structure protection specialists to determine which homes they could defend in Yarnell and Peeples Valley, then notified the sheriff's office to prepare for evacuations.[6]

But orders the team had made for resources earlier in the day were already going unfilled. Dispatch declined to send a heavy air tanker and helicopter from Prescott on Saturday afternoon, June 29, due to severe weather. One plane was grounded by an oil leak. Another sat on the tarmac in Wickenburg, just down the road from Yarnell, while dispatchers debated which fire needed it most. A VLAT was available in Albuquerque and would be unaffected by the weather, but incident command in Yarnell declined it, believing the steep terrain and coming darkness would keep it from delivering retardant effectively.

Just after 6 p.m. Russ Shumate contacted Charlie Havel at the Arizona Interagency Dispatch Center in hopes of getting two hotshot crews to Yarnell at 6 a.m. the following morning. Havel told Shumate that the fire was "sitting low" on the priority list, but he would do what he could.

At 6:21 Havel asked the Southwest Coordination Center, the Albuquerque branch of the National Incident Management System, to send two Type 1 crews to Yarnell. Four minutes later the SWCC replied that they could send only one crew, the Arizona-based Blue Ridge Hotshots.

"That will be the only [hotshot crew] I have for tomorrow, though," the SWCC stated.

At 8:10 the Arizona dispatcher contacted the SWCC to request a specific crew. "Placing order for Granite Mountain IHC."

"Can't accept assignment," the SWCC responded three minutes later.

Arizona dispatch did not relent in its efforts to get another hotshot crew. "We have pushed orders for another Type 1 crew," they reported to the SWCC at 8:49.

Havel notified Arizona fire managers and dispatchers that he had "e-mailed a resource order to Eric Marsh for Granite Mountain Crew."[7]

That seemed like a bypass of the normal dispatch system, but it wasn't surprising that Marsh accepted the assignment. The fire was in Granite Mountain's backyard. Many of the hotshots knew people who lived around Yarnell and would do anything they could to help them. And Marsh went way back with many members of the incident command he'd be working with there.

Around 10:30 Shumate called Darrell Willis, Marsh's boss, and asked him to oversee firefighters protecting homes in Peeples Valley. Willis's crews scouted the town overnight to prepare for the fire's arrival. They found Double Bar A Ranch, just outside the town, surrounded by grass and scrub up to eight feet high. The buildings there would be hard to save, and the fire could easily move from the ranch through the Model Creek subdivision and into the town.

On the other side of the fire, structure protection specialists in Yarnell determined that at least half of the homes there couldn't be protected at all. Most of them had no defensible space and had been built without any flame-resistant materials. Nonetheless, firefighters began positioning themselves to make a stand.

THE GRANITE MOUNTAIN HOTSHOTS ARRIVED in Yarnell before 8 a.m. Sunday. From the fire station they could see the burning ridge three miles west of the highway. Todd Abel, an operations section chief, assigned them to anchor a fire line at the south end of the blaze and then build a line to keep the fire from coming into town. Eric Marsh, who had arrived earlier, sat in on a 7 a.m. briefing with the outgoing and incoming incident commands. He was assigned to supervise Division

Alpha of the fire. His crew was assigned to that division, but his responsibilities in the management team meant he once again wouldn't be directly in charge of them. Jesse Steed, the crew's captain, would supervise the Granite Mountain Hotshots.

The crew didn't get a briefing or a current map of the fire, but Marsh joined the incident command team to look over a map of the area where it was burning on an iPad.[8] They noted a small, tan blotch amid the green of the chaparral — Boulder Springs Ranch, where the owners had cleared so much vegetation, the dirt could be seen from space. They rode toward their assignment with Gary Cordes, who was overseeing structure protection in Yarnell and reminded them that the ranch was a "bombproof" safety zone. But, he also noted, the easiest place to escape the blaze was the black, where the fire had already burned away most of the fuel.[9]

The new command team, led by Roy Hall, who had nearly 40 years of experience fighting wildfires, took charge of the fire at 10:22 a.m., setting up shop in the Model Creek School in Peeples Valley. But like the day before, the firefighters were slow getting up to speed. Command team members were arriving as they could, and there just weren't enough of them to manage a situation that was growing more complex by the minute.

The communications manager wouldn't arrive until midday, and even then firefighters complained that there was nobody available to "clone" their radios onto the frequencies assigned by incident command. Many radios didn't have "tone guards," without which other radios ignored their transmissions.[10] Hall said he didn't see the communications manager at all on Sunday.

The base camp manager arrived with trailers of equipment later that afternoon,[11] but by then the command post was on the verge of being overrun by the fire and the incident commander was preparing to evacuate the school. Resources remained hard to come by. A large helicopter and two heavy air tankers headed to Yarnell were diverted to other fires. A military C-130 equipped with a MAFFS loaded up with retardant in Colorado but never took off due to the weather. The dispatch center ordered two more planes, but only one of those arrived.[12]

By noon the fire had burned at least 1,000 acres. It was growing fast and launching an increasing number of embers to start spot fires. During the afternoon several helicopters, four SEATs, and three heavy air tankers worked the blaze from the air, soon to be joined by both of the nation's VLATs. That still wasn't enough. The retardant was hardly slowing the fire. The Southwest Coordination Center asked for six more large air tankers from the National Interagency Fire Center in Idaho.

"Very limited availability of [air tankers] with increasing activity in the western states," came the response. "Unable to fill at this time."

Marsh and his crew parked their buggies on Sesame Street, a private road with a few houses, a little after 9 a.m. They marched up the mountain for 45 minutes. An hour later the Blue Ridge Hotshots arrived at the fire.

JOY COLLURA AND SONNY "TEX" GILLIGAN were the odd couple of hiking around Yarnell. The 40-year-old Collura, who had found hiking as she recovered from health problems that included brain surgery, and the 70-year-old Tex, who was struggling with alcohol and the death of his son, often set up camp for weeks or slept in caves or mine adits as they traipsed through Arizona's mountains. Collura's husband encouraged his wife's adventures because they seemed to improve her health.

"We're like an old married couple," Collura joked about her relationship with Gilligan. "I'm married, and he's old."

Around 8:30 Sunday morning they hiked up Yarnell Hill behind Boulder Springs Ranch and ran into Eric Marsh hiking up the mountain from the east. "What's the best way up?" he asked. Gilligan pointed out the jeep road nearby.

As morning passed into afternoon, they ran into him again, this time just below where the lightning strike had started the fire. The fire and retardant drops from aircraft were picking up.

"You two are going to have to get out of here soon," Marsh warned.

Later, they passed the Granite Mountain Hotshots as the crew hiked up the two-track road. Collura took a few photos of them, but they didn't chat. To Gilligan the crew looked spent. "It was like a death march," he said.

Collura started to climb down a steep gully toward Boulder Springs Ranch, which they could see below them. Tex, an old desert rat who knew what Collura was getting into, headed her off. "That way's gonna get you killed," he said, explaining that the chaparral and boulders would be so thick it would take hours to find a way through. And if the fire got in there, the whole canyon would blow up.

They hiked out via an easier route and thought little more about the firefighters they had seen.

Marsh assigned most of his crew to burn out the scrub along the two-track road in hopes of corralling the eastern edge of the blaze. Then he led Brendan McDonough and a couple of other hotshots up the mountain to build an anchor point—a position they cleared completely and tied into the black to start their fire line.[13]

At 11:36 and again at 11:45, air tankers bombed the fires the crew had set.[14] Marsh was peeved that they could no longer burn out a fuel break, but scrapping plans and changing tactics was part of his job.

At 11:54 the superintendent and the captain of the Blue Ridge Hotshots, Brian Frisby and Rogers Trueheart Brown, arrived on an ATV at the anchor point Marsh's crew had built. The Blue Ridge crew had arrived at the fire before the supervisor of their division, so they lacked direction, and the safety officer wouldn't be there until late in the day. With radio communications increasingly difficult and the two hotshot crews never having met to discuss tactics, the leaders decided to meet face-to-face. The Blue Ridge Hotshots were working less than a mile away, putting in a line with a bulldozer along the foot of the mountain. With his plan to conduct a burnout thwarted, Marsh ordered his crew to dig a line along the eastern flank of the fire—a direct attack that would put them close to the flames. They'd need a lookout. Donut had been sick for a couple days before the crew had come to Yarnell, so he drew that lighter-duty assignment. He headed down the mountain, found a knoll near a road grader on the ranchland below, and called Jesse Steed, the captain, to let him know his position and that he had his eyes on them. Several hundred yards to the north he identified a trigger point that, if the fire reached it, would prompt his own evacuation.

Marsh, in charge of Division Alpha, the southwest end of the fire, also split off from his crew. The northeast part of the blaze was in Division Zulu and put under the supervision of Rance Marquez, a BLM firefighter. The boundaries between the divisions, however, were unclear, and Marsh and Marquez argued about whose turf was where, leading to even more confusion.[15]

Granite Mountain was the only crew assigned to Marsh's division, so he was still managing only his own people, but overseeing the entire area meant he wouldn't necessarily be with them. He hiked alone up the mountain to formulate a plan.

AIRCRAFT REPORTED THAT THE EASTERN FLANK of the fire was active and moving toward Yarnell. At 1:50 the Yavapai County Sheriff's Office initiated a pre-evacuation notice for the town. Twelve minutes later Brian Klimowski with the National Weather Service in Flagstaff warned of likely thunderstorms on the east side of the fire, with wind gusts up to 45 miles per hour.[16] The Incident Management Team's fire behavior analyst relayed the warning to the division supervisors, including Marsh. At 3:26 the National Weather Service sent another warning of thunderstorms and winds from the north-northeast gusting up to 50 miles per hour.

Rain that evaporates before it hits the ground is called virga. It hangs below thunderstorms like a swinging horse's tail. While the deluge doesn't reach the earth, the air it evaporates into becomes cold and heavy, plunging like a waterfall void of water. The downdraft hits the ground in hard gusts and splashes out in directions that are notoriously hard to predict. The wave from the air's collision with the ground, known as an outflow boundary, can drive a fire back, push it forward, or turn it in an unexpected direction.

Todd Abel confirmed with Marsh that he had received the weather reports and could see the clouds building over the smoke.

"The winds are getting squirrelly up here," Marsh reported.[17]

At 3:50 Marsh called, "I'm trying to work my way off the top."

"Just keep me updated," Abel said. "You guys hunker and be safe and we'll get some air support down there ASAP."

The Granite Mountain crew had stopped for lunch where the fire had burned away most of the vegetation about a quarter mile below where Marsh had been scouting the fire. From the safety of the black they watched the weather drive the fire hard.

Donut, the lookout, saw the fire hit his trigger point and let Jesse Steed know he was moving from his post.

"OK, cool," Steed responded.

"I've got eyes on you and the fire," he radioed to Donut a few minutes later. "It's making a good push."

The fire was growing so ferociously that Donut worried it might overrun him as he hiked out. He was going to call for a rescue ride from the Blue Ridge Hotshots when Frisby appeared on his ATV. Donut climbed aboard but he couldn't retreat from the area yet. Leaders of the two hotshot crews had realized that Granite Mountain's buggies were in the path of the fire. Blue Ridge Hotshots helped Donut move the trucks, first to Shrine Road, where the dozer line the Blue Ridge crew was building started, and then, as the fire exploded, in a convoy with their own vehicles to the Ranch House Restaurant at the south end of U.S. 89. On their way out they warned other crews who had made their way up washes to engage the blaze to evacuate before it was too late. Then they joined the retreating firefighters and residents who watched the growing firestorm from the parking lot.

WHEN THE THUNDERSTORMS ARRIVED in Peeples Valley, Darrell Willis had been working there for nearly 18 hours straight.

"When I arrived, around midnight on the 29th, we were so far behind the eight ball that there was no catching up," he told me.

Willis had no maps of the area when he took charge of protecting homes on the north end of the fire. The firefighters preparing the town overnight put in orders for trucks, planes, and crews, but they knew the paperwork wouldn't get finished before daybreak and help wouldn't arrive until hours after that, if it was available. The only thing they were certain would get there was the fire. "I knew that the fire was going to come into Peeples Valley," Willis said.

It came earlier than they expected, charging toward Double Bar A Ranch before 10:30. "That's really unusual fire behavior for that early in the morning," he said.

When the thunderstorms hit that afternoon, Peeples Valley seemed certain to burn.

By four that afternoon Conrad Jackson (the high school teacher turned firefighter), Willis, and several others were lined up along a road between the ranch and the oncoming flames. In the town behind the ranch, some homes stood a chance, but others had dry and oily chaparral so close that a fire could just run right onto the roof.

"There are some nice properties in there, but nobody's done any defensible space," Willis said. "If we can keep it on the road and not let it cross . . . to where the homes are on this side, we'll be good. If we don't, we're going to lose them."

Among the buildings about to be overrun was the Model Creek School, where incident command was set up. With the flames less than 200 yards away, Roy Hall, the incident commander, prepared to evacuate the command post and ordered all vehicles moved to the north side of the school, away from the coming flame front. Some of the firefighters raced to clear brush and flammable materials from around structures. Others hosed down encroaching flames or wet down houses. Willis decided that a burnout along the road was their only chance to save the town. "It's the last stand," he said.

By charring the chaparral along the road, they could starve the fire of fuel. It was a dangerous tactic — the backfire could live up to its name and spread into the town. After crews lit the brush along the road, it looked like that would happen. "We're just constantly losing it," Willis said. "We [can't] keep up with the spots."

Firefighters chased down the spot fires and slop-over where the burnout crossed the road. Whether from the wildfire or the burnout they had lit to fight it, they were about to be overrun. Willis ordered the crews to retreat. Jackson and his partner, Mark Matthews, pulled back from the ranch to a rise where they could watch the fire burning into the town.

"And then, all of a sudden," Willis said, "the wind shifts."

At about 3:50 the winds that had been blowing the fire to the northeast turned it hard to the southeast, shifting the fire away from Peeples Valley. The firefighters breathed shocked sighs of relief. Jackson and Matthews cheered.

But the winds that had saved one town doomed another. And the outflow boundary from the thunderstorm hadn't yet arrived.

• • •

THE GRANITE MOUNTAIN HOTSHOTS, high in the black, were as safe from the flames as they could hope to be, but they were exposed to the forces of the changing weather and had a panoramic view of the sudden turn of the fire it drove. Many of them were in touch with their families back in Prescott.

Grant McKee's fiancée, Leah Fine, sent him a note. "Wish I could come kidnap you and take you away," she texted him.

"Please do," he wrote back. "I dare you."

She reminded him to wear sunscreen to keep from getting burned.

Wade Parker sent his mother a text with a photo. "This thing is runnin straight for yarnel jus starting evac," he wrote. "You can see fire on left town on right."

Parker's photo showed a line of flame with a billowing plume of smoke leaning toward Yarnell.

Scott Norris also texted a photo to his mother. "This thing is running at Yarnell!!!" he wrote.

Sometime between 3:45 and 4:00, Paul Musser, one of the operations chiefs on the fire, requested that the Granite Mountain Hotshots make their way to Yarnell to assist in protecting structures there. Either Marsh or Steed refused this request and suggested that the Blue Ridge Hotshots were closer.[18] Marsh was also part of a short conversation about trying to backburn an area next to the dozer line the Blue Ridge crew had cut that morning, but that wasn't an option either. Still, Marsh was looking for opportunities to help.

Chris MacKenzie made a short video that panned the fire and captured Marsh speaking to Steed over the radio. "I knew this was coming when I called you and asked what your comfort level was," Marsh said. "I could just feel it, you know?"[19]

He could have been referring to the weather, the pressure for them to get to Yarnell, or something else.

A few minutes later Marsh called the Blue Ridge crew. "I want to pass on that we're going to make our way to our escape route," he said.

"You guys are in the black, correct?" Frisby, the Blue Ridge super, asked.

"Yeah," Marsh answered, "we're picking our way through the black . . . going out toward the ranch."[20]

One of the photos from MacKenzie's phone, on which he also recorded his videos, appears to show some of the crew's sawyers headed south at 3:52. The rest of the crew started down about 12 minutes later.

The Blue Ridge superintendent assumed that Marsh and the Granite Mountain crew were heading toward one of the ranches to the north, through already burnt terrain that provided their safest route off the mountain. But that route would take them away from the fire and Yarnell. In reality, they headed south, back along the two-track road where they had conducted their earlier burnouts. To the east, a little more than half a mile below the two-track, was Boulder Springs Ranch, the place they had noted as a "bombproof" safety zone that morning.

But between the safety of the black they were abandoning and that of the ranch was a steep canyon choked with boulders and thick chaparral.

31

TRIGGER POINTS

IN YARNELL, CORDES ESTABLISHED three trigger points for evacuations.[1] The first was a peaked ridge about a mile north of town. If the fire reached it, the Yavapai County Sheriff's Office would give residents an hour to leave. If the fire reached the second trigger, a hill south of the first, firefighters would retreat from the wildlands and pull back from the northwestern edge of town. And if it reached the third, a ridgeline near the edge of town, everyone would get out.

Since the day before, the Yarnell Fire District's website had warned residents to be prepared to evacuate, but had presented the possibility as remote. The fire was moving away from the town, the website noted, but the potential for an evacuation "must always be considered." By 3:30 Sunday morning, the fire was about a mile and a half from the town line. It wasn't until late that morning that the sheriff's office contacted residents via telephone, text, and email.

"Residents receiving this message should be prepared to evacuate due to the Yarnell Fire. Collect your valuables and make arrangements for livestock transportation. If evacuation becomes mandatory, we will send out a second notice. You will have one hour upon receipt to evacuate."

It was more difficult to get the word out than in the past, though. Yarnell had disbanded its volunteer Community Emergency Response Team. The town's emergency siren had been stolen, then recovered, but it was still out of commission because, according to some residents, rats had chewed through its wires. Repairing it hadn't been a priority, due in part to noise complaints whenever it was tested.[2]

Written evacuation plans are standard procedure in wildfires, but nothing was put on paper in Yarnell on Sunday. To some degree that was due to the changeover of incident management, commander Roy Hall later reported.[3] Through the early afternoon leaders held off ordering an evacuation for fear of starting an unnecessary panic. Many in Yarnell reported not receiving the automated phone, text, or email evacuation notice. Only landlines with listed phone numbers were automatically included in the system, and many residents with cellphones or unlisted numbers didn't realize they had to sign up for the service.[4]

At 2:26 the sheriff's office advised residents that they were on a four-hour standby to evacuate and that they would receive another notice when and if the evacuation became mandatory. But the fire was already nearing the trigger point for the one-hour evacuation.

When the winds turned, shortly after 3:30 p.m., 41-mile-per-hour gusts blew the fire fast past that landmark. Incident command requested an immediate evacuation of Yarnell, and the sheriff's office sent an emergency alert at 4:08. Heavy ash and embers fell on the firefighters near the Shrine of St. Joseph. The blaze traveled more than a mile in 15 minutes and hit its second trigger point at 4:22. By the time firefighters around the shrine began evacuating, embers were starting fires in town. After 15 minutes more, rather than the hour that firefighters had planned on, flames overran the third trigger point.[5]

Officers raced through the town's byzantine roads with sirens wailing, shouting through bullhorns for residents to get out. Less than 15 percent of the area's homes had received automated evacuation notifications and the mandatory evacuation was delayed some 20 minutes while dispatchers gathered information to map the alert area. Some were confused by the contradictory messages. A number weren't aware that the fire was upon them until neighbors or officials knocked on their doors.

Bryan Smith, 63 and disabled, didn't know he needed to evacuate until he saw the fire about to climb the steps of his home in the Glen Ilah subdivision. He hosed down the stairs, then roused his 85-year-old cousin, Pearl Moore. Smith's wife had their car, so he and Moore starting walking out of town as the trees ignited around them and embers fell onto their hair and bare skin. Both sides of the road were in flames, and they could hear tires popping on vehicles, so they walked in the middle of the road. When Moore could walk no further, she begged Smith to save himself.

"Help me, Jesus," she told him. "I'm ready to go."

Smith laid her down on the side of the smoky street and ran to get help. He saw a truck with flashing lights around the corner. Gary Cordes, who was overseeing the firefighters in charge of structure protection in Yarnell, had given up on saving homes but was still trying to save their residents. Moore was the last one he rescued.

Another elderly couple, Ruth and Bob Hart, were ready for bed when they saw flames in their garden. They crashed their car while trying to escape and walked out of the burning town in their pajamas, bathrobes, and slippers.

By nightfall 127 homes had burned. Smith's still stood. The Harts' had burned.

A FIRE OFFICIAL TOLD INVESTIGATORS later that the evacuation trigger points were off by more than 50 percent. Firefighters and residents racing out of the burning Glen Ilah subdivision were convinced that people had died there. It took more than two days for officials to confirm that everyone in the town had survived.

32

NINETEEN

AS YARNELL BURNED, ALREADY DIFFICULT communications fell into chaos. Firefighters who couldn't get through on radios used cellphone calls and text messages. Few heard anything from the Granite Mountain Hotshots for more than half an hour, but tapes of radio traffic recorded several transmissions from Eric Marsh amid the static, wind, and chatter. He reported that the hotshots were headed south toward Yarnell rather than north into the safety of the black.

The pilot watching over the blaze reached his maximum allowed flying time as the crew descended. He left, and another plane, *Bravo 33*, which had been guiding retardant drops, put eyes on the fire.

That plane's pilot asked about the Granite Mountain crew after hearing Marsh's talk of moving down an escape route. "I heard a crew in a safety zone," *Bravo 33* reported to Todd Abel. "Do we need to call a time out?"[1]

"No, they're in a good place, and it's Granite Mountain," Abel replied. *Bravo 33* then called Marsh. "Is everything OK?" the pilot asked.

"Yeah, we're just moving," Marsh responded.

The crew followed the two-track road until they came to the top of a box canyon. They could see the Boulder Springs Ranch safety zone at the bottom. It was just over half a mile away, but the steep bowl in between was dense with boulders and thick with brush. They took a sharp left turn off the road to drop into the canyon. As they descended, the canyon wall to their north blocked their view of the fire.

"Division Alpha, what's your status right now?" one voice asked

Marsh in radio traffic recorded in the background of an air tanker's communications.[2]

"The guys, ah, Granite, is making their way down the escape route from this morning. It's south, mid-slope, cut vertical."

"Copy," another voice responds, perhaps acknowledging that the crew's ultimate plan was to fight the fire in town. "Working their way down into structures."

The crew was cutting its way down the canyon toward the ranch. Marsh may have been leading the sawyers to cut their way through the dense chaparral, but the rest of the exchange appears to indicate that Marsh had already reached the ranch house.

The first voice asked Marsh, "On the escape route with Granite Mountain right now?"

Marsh responded, in a radio transmission that was almost indecipherable when it was discovered. "Nah, I'm at the house where we're gonna jump out at," he appears to respond.

At least that's what the former firefighter who discovered the recorded radio call believes he said. Others who have heard the recordings dispute that. The contents of the nearly inaudible recordings of Marsh's transmissions during the crew's descent from Yarnell Hill, which are critical pieces of evidence that may establish that the leader of the Granite Mountain Hotshots had reached the safety zone and then returned to the fire to assist his crew, have been subject to heated debate.

At 4:30 another voice radioed Marsh. "Copy . . . coming down and appreciate if you could go a little faster but you're the supervisor."

"They're coming from the heel of the fire," Marsh reported, another hint that he may have moved ahead of the crew.

About that time the outflow boundary of the downdrafts hit the south end of the fire, splashing it onto the mountains and into Yarnell.[3] A pyrocumulus cloud rose 40,000 feet into the sky. The Granite Mountain crew was near the floor of the bowl they were descending when the smoke column rose like a giant cobra, then struck at the Weaver range. The fire that the weather had pushed southeast toward Yarnell made another hard turn, this time to the southwest — into the canyon the hotshots were descending.

The weather alerts led the crew to expect the turn they saw the fire make when they were in the black, but they probably didn't anticipate the second one. The natural chimney of the canyon magnified the wind like a giant nozzle. The fire that they couldn't see during their descent was suddenly huge and on top of them. They could never outrun it up the rocky, brush-choked slope they had just descended or make their way past the flames blocking the canyon.

At 4:37 Marsh saw *Bravo 33* fly over the canyon as it guided an air tanker in for a retardant drop. "That's exactly what we're looking for," he called to them over the radio. "That's where we want the retardant."[4]

It wasn't clear that he needed the retardant drop to save him and his men.

At 4:39, as the aircraft circled back over them, a static-filled and overmodulated call from someone on the crew interrupted their radio communications again. This time there urgency in the voice. "Breaking in on Arizona 16, Granite Mountain Hotshots. We are in front of the flaming front."

The radio calls are muffled by the wind and static.

". . . Granite Mountain, air attack, how do you read?" a crewmember called to the plane. "Air Attack, Granite Mountain 7, how do you copy me? Air Attack, Granite Mountain 7!"

With the airwaves filled with radio traffic from the firefighters trying to save Yarnell, the pilot at first thought the broken-up call was coming from there, rather than the hotshot crew that, the last he knew, was in a safety zone.

"Division Alpha with Granite Mountain," Marsh eventually called. "Yeah, I'm here with Granite Mountain Hotshots. Our escape route has been cut off."

Chainsaws are running in the background of the transmission.

While few people had heard Marsh's radio calls over the previous 33 minutes, the ones the crew made to air attack were on a channel monitored by many of the firefighters. Two firefighters watching the blowup captured the calls on a helmet cam.

"Is Granite Mountain still in there?" one asked.

"Well, they're in a safety zone," the other responded. "In the black."

"Air attack, Granite Mountain 7, how do you copy me?" came a Granite Mountain Hotshot's voice over the radio.

"I hear saws running," one of the listening firefighters said. "That's not good."

"Not when they're in a safety zone."

"Air attack, Granite Mountain 7!" came another transmission.

"This ain't good," one of the firefighters said.

"No, he's screaming."

It was hard to imagine a worse site to deploy their shelters, which are designed to be held snugly to level dirt with nothing that can burn nearby. The thick, oily brush and boulders blocking the canyon not only magnified the intensity of the fire but left the firefighters with few appropriate spots to lay down their foil cocoons. As the natural chimney sucked the flames uphill, winds greater than 50 miles per hour pushed them from behind. The crew had just minutes to clear the site for their shelters.

"We are preparing a deployment site and we are burning out around ourselves in the brush and I'll give you a call when we are under the sh . . . the shelters."

"Okay, copy that," *Bravo 33* responded. "So you're on the south side of the fire, then?"

"AFFIRM!" Marsh shouted.

THE PILOT MADE SEVEN CALLS to the hotshots over the next four minutes. Below him the outflow winds blew 70-foot flames sideways into the canyon. An air tanker circled nearby, ready to drop retardant on the hotshots as soon as it found them. Despite the crew's last calls, nobody knew where they were.

In the Ranch House Restaurant parking lot, where firefighters had retreated from the conflagration in Yarnell, the word "deployed" on radios turned heads away from the burning town. Brian Frisby and Rogers Trueheart Brown, the Blue Ridge Hotshots' superintendent and captain, headed back into the fire on ATVs, carrying a couple of bottles of oxygen, a medical kit, and a backboard. With the fire front spreading almost parallel to the highway, they found their way back to Yarnell Hill blocked by exploding propane tanks, fallen power lines, and fire. They urged residents who still hadn't evacuated to leave, then stopped at a wall of flame on Shrine Road. Frisby announced, "Fuck it, let's go for it," and they charged through the heat to the black, safe ground beyond. They made it back to where they had last seen Granite Mountain and then up Yarnell Hill to the anchor point the crew had cut that morning, but they found no sign of Marsh's crew.[5]

In Peeples Valley firefighters were catching their breath after the fire turned away from that town. Conrad Jackson was still with his fire truck on the ridge when he got a call.

"Have you heard something about Granite Mountain?" the caller asked. "We're hearing something bad happened on the fire."

Mountains hindered his radio's reception, but he eventually caught a snippet indicating that a command had been set up for a new incident tied to the hotshots. Jackson had been on fires where people deployed before, and all of those firefighters had walked away with a few burns. He kept listening, then heard "18 accounted for, no injuries," and breathed a sigh of relief.

He texted his buddies what he had heard.

"I don't think you're right," one responded. "They've got some pretty definitive sources up here that are saying there's been a deployment and there may be fatalities."

A helicopter had been prepared to launch since Marsh had first reported the hotshots were in trouble, but smoke and wind kept them grounded until 35 minutes after his last radio call. They spent an hour flying over the area where the hotshots had been working, then swooped toward Boulder Springs Ranch. Three hundred yards west of the ranch, they spotted fire shelters through the smoke and landed. Eric Tarr, a police officer and paramedic with the Arizona Department of Public Safety, hiked across the crusty, burnt ground and made his way from shelter to shelter to take each man's pulse.

WHEN BRENDAN MCDONOUGH HEARD the news about his crew, he sat alone in one of Granite Mountain's buggies in the restaurant parking lot until the phones some other members of the crew had left in the truck started ringing with family and friends trying to reach them.

In Prescott, J. C. Trujillo, the old bronco rider, was helping a crew set up a tent for the opening of the Frontier Days rodeo the next day when a hard wind nearly blew it away. He looked at the smoke in the distance and wondered what the weather was doing to the fire.

The storm blew clouds over the city, and rain sprinkled on Claire Caldwell, who was out watering her garden. She texted her husband. "I hope this rain is helping you guys out," she wrote. "Maybe you'll get to come home tonight."

She went in, had dinner, and was watching a movie when she heard someone yelling in her living room. A friend who worked for the Prescott Hotshots was standing there. Claire knew he was supposed to be on a fire.

"Granite Mountain had a burnover today," he told her.

Claire collapsed. She knew what that meant.

He drove by one of the city's fire stations, and then to Prescott High School, where the families of the firefighters were gathering. She saw the men and women from her wedding. Big firefighters were weeping.

Although the crew had perished together in seconds, to Claire it seemed like they were falling one at a time. Claire kept asking Amanda Marsh what was happening.

"We lost Clayton," Amanda responded, and Claire went to comfort his wife.

When Claire saw Amanda again, she begged for more information.

"We know we lost Grant," she said.

Grant was Robert's cousin. Claire had worked with him. The tragedy was closing in on her, but she kept expecting the buggies to show up and the rest of the crew to get out. Finally, she grabbed an official.

"Is my fucking husband alive?" she yelled.

He asked her name, checked a list.

"No," he told her.

Grief therapists tried to hold her, but she shook them off and ran into a stand of trees by the school. She wanted to find a stream to lie down in.

"Now we really are the Granite Mountain Girls," she thought.

IN PEEPLES VALLEY, CONRAD JACKSON got another call.

"They're all dead."

Jackson was numb when he arrived at the fire camp. Nobody would look at him as he walked through. Men had died doing what he had taught them to do in high school.

Then Donut stepped out from behind a fire truck.

"Oh, it's all bullshit!" Jackson thought for a second. "There's my guy!"

But of a crew of 20 that had marched up the mountain, he realized there was only one left.

He and Donut buried their heads in each other's shoulders and sobbed.

"I thought I lost all my boys," Jackson whispered to him.

33

BLOWBACK

WHEN I ARRIVED IN ARIZONA the day after the Granite Mountain Hotshots died, the world seemed to embrace Prescott. Tributes hung on the fence around Station 7, eventually arriving from as far away as Australia. In the coming weeks thousands of items would cover every inch of the chain-link fence and spill out across the ground in front of it.

The crew's bodies returned from Phoenix through Yarnell and then back to Prescott in 19 white hearses, behind a convoy of wildland buggies, fire trucks, and motorcycles. When the escorts pulled up on Whiskey Row, the crowds along the street cheered, but with the appearance of the first white hearse, they fell silent.

The impromptu memorial gathering that I attended in the Embry-Riddle Aeronautical University gymnasium, exactly 24 hours after the hotshots had deployed their shelters, was followed by more formal services. One filled Prescott High School's football field and bleachers, where residents and family members sat in their shirtsleeves listening to guitars and ministers as the setting sun and smoke turned the sky red. A week later Vice President Joe Biden joined Arizona's governor, senators, and dozens of other officials at the arena where Wade Parker had had tickets to see a Christian rock band on the weekend he and his crew perished. They addressed the crowd that filled the 5,100-seat arena to capacity from a stage overlooking photos of the hotshots, 19 shovels and Pulaskis, and 17 hard hats. The day before the service more than 100 bagpipers and drummers had marched up the timeline in front of the Yavapai County Courthouse — the largest gathering of its kind since 9/11, marking the largest loss of life of U.S. firefighters since the terrorist attacks. Just a single piper marched through the arena during the memorial.

At the podium Donut, the lone survivor of the crew, shared his first words with the public since the tragedy, a reading of the Hotshot Prayer.

> *When I am called to duty, Lord . . .*
> *To fight the roaring blaze . . .*
> *Please keep me safe and strong . . .*
> *I may be here for days.*
> *Be with my fellow crewmembers . . .*
> *As we hike up to the top.*
> *Help us cut enough line . . .*
> *For this blaze to stop.*
> *Let my skills and hands . . .*
> *Be firm and quick.*
> *Let me find those safety zones . . .*
> *As we hit and lick.*
> *For if this day on the line . . .*
> *I should answer death's call . . .*
> *Lord, bless my hot shot Crew . . .*
> *My family, one and all.*

Before he returned to his seat, he choked back tears.

"I miss my brothers," he said.

Outside, more than 20,000 people filled the parking lot in the 100-degree temperature to watch the service on giant television screens. When it ended, hundreds of firefighters lined the drive into the arena and came to attention as the remaining members of the Prescott Fire Department marched out in their green pants and black T-shirts in front of buses that carried hundreds of family members and friends of the fallen firefighters.

By the box office, posters of the 19 fallen firefighters adorned with purple ribbons stood behind the giant bronze sculpture of a firefighter carrying a Pulaski. Once part of the Wildland Firefighters Monument that's adjacent to the National Interagency Fire Center in Boise, Idaho, the sculpture would now reside in Prescott. Another poster nearby advertised the next event at the arena — the Doomsday Prepper and Survivalist Expo. Across the street, bikers in leather adorned with skulls and flames handed out doughnuts.

Across the country, the disaster spawned hundreds of fund-raisers. Rallies of Harley-Davidson riders; Cub Scout car washes; CrossFit competitions; concerts; sales of T-shirts, stickers, and jewelry honoring the hotshots; and firehouse "fill the boot" collections. Larger gifts arrived from Arizona's 100 Club, which supports the families of fallen police officers and firefighters, and the Wildland Firefighter Foundation, which received more than $2 million in donations after the tragedy and transported the statue of the firefighter that now stood in front of the arena. In all, these fund-raising efforts raised some $13 million by the end of the year, with more than $11 million earmarked for the hotshots' families and much of the rest going to the Yarnell community.[1]

But by then the tragedy that had brought the community together was tearing it apart. The hotshot crew that had seemed such a clever solution to the city's wildfire hazard ended up surrounded by burnt bridges.

The hotshots left 10 wives, 3 fiancées, and 16 children — three of them unborn. While all of the families would receive at least $328,000 in federal Public Safety Officers' Benefits, only the families of the six full-time hotshots qualified for benefits such as pensions, health insurance, and life insurance. Widows who were unaware before the tragedy that they were ineligible for the city's benefits were outraged. Their husbands had worked the same number of hours during the fire season as the full-time hotshots had.

Reclassifying the seasonal employees as permanent would violate state law, the city argued. It would also devastate the city's finances. An assessment of the cost of the hotshots' benefits determined that if the seasonal firefighters received the benefits given to the permanent employees, it would cost the city at least $51 million over 60 years. As a onetime payment, the liability would add up to three times the Prescott Fire Department's budget.[2] Some city officials said they were unaware that Prescott even had a hotshot crew and would never have approved it had they known the potential costs.

Juliann Ashcraft, Andrew's slender, demure wife, and the mother of his four children, was the first to speak out and became the face of the aggrieved families.

"As shocked as I was that my husband went to work and never came home, I'm equally shocked in how the city has treated our family since then," she told CBS News. ". . . I said to [city officials], 'My husband was a full-time employee, he went to work full-time for you.' And their response to me was, 'Perhaps there was a communication issue in your marriage.'"[3]

She held a press conference on the Yavapai County Courthouse steps with other widows and was the first of them to threaten a lawsuit against the city.

"Quite literally, my bills are being paid by the good people of the world who are giving donations, because the city of Prescott isn't doing anything for us," she said. "Now I have four kids and myself, and I don't know what I'm going to do."

The city pushed back.

"She's a neat little lady . . . but the money took hold in this situation real fast," Prescott mayor Marlin Kuykendall told the *Prescott Daily Courier* barely a month after the hotshots' deaths. "This is big bucks, when it's all over, big bucks. And the money seems to be leading some of the battle over the city's participation."[4]

Ten months after the disaster, Prescott's Public Safety Personnel Retirement System board voted to award full benefits to Ashcraft's family. The city council appealed the board's decision, claiming it was based more on emotion than the facts of the case, but after another eight months, a superior court judge upheld the board's decision. The board eventually did award full benefits to Ashcraft's family, as well as to those of Sean Misner and William Warneke. The city declined to appeal again.[5]

In the meantime, 12 families of the firefighters jointly sued the state, asking for $10 million for each firefighter's wife, $7.5 million for each surviving child, and $5 million for each of the parents filing suit. In Yarnell more than 160 people who lost property sued Arizona, claiming that negligence in the firefighting operation led to the loss of their homes as well as the deaths of the hotshots.[6] While most Yarnell residents felt nothing but sadness about the men who had perished trying to save their town, and sympathy for their families, conspiracy theories abounded.

In a bar outside of town, I sat with one resident who told me that the firefighters had let the fire burn intentionally. "That's how they get more funding," he argued. "By letting the fire get big."

In other corners, the legal battles of the families and homeowners alienated community members who noted that they had raised millions of dollars for the firefighters' families and for Yarnell residents who had lost their homes and that the lawsuits and fights for benefits were hurting the city and the state. For the hotshots' widows who fought for benefits for their families, accusations of being greedy and opportunistic added to their grief. Juliann Ashcraft moved her family away from Prescott to escape the stinging criticism.

They weren't the only ones angered by the response to the tragedy.

In September, two months after the Yarnell Hill Fire, the Serious Accident Investigation Team, organized by the Arizona State Forestry Division but made up primarily of firefighters from outside the state, released its 116-page report.[7] While filled with documents and details about the firefight, right down to the condition of each of the fallen hotshots' fire shelters, the report was cripplingly incomplete, due in large part to the investigators' two-month deadline, which left them with only a fraction of the time they needed to do a more thorough job. The team concluded that the actions of incident management were "reasonable" and "found no indication of negligence, reckless actions, or violations of policy or protocol." In short, nobody had done anything wrong. Outraged firefighters across the country argued that the failure to attribute errors all but guaranteed more death and destruction like that on Yarnell Hill.

In early December the Arizona Division of Occupational Safety and Health (ADOSH) released a report produced by Wildland Fire Associates, a private wildfire investigating company. That report concluded that the Arizona Department of Forestry and Fire Management, which oversaw both the firefighting effort and the first report about the tragedy, prioritized protecting private property over firefighter safety. ADOSH fined the Arizona State Forestry Division $559,000 for "serious, willful" safety violations.[8]

Both investigations resulted in more questions than answers. The lack of survivors or witnesses to the hotshots' last hours was compounded by a lack of cooperation between federal and state authorities, preventing most of the people who last saw the crew alive from being interviewed in detail by ADOSH investigators.

Immediately after the tragedy, the U.S. Forest Service sequestered the Blue Ridge Hotshots. The federal crew had been working closest to the Granite Mountain Hotshots on Yarnell Hill and were in a better position to see what led up to their deaths than anyone else on the fire. The feds allowed the Serious Accident Investigation Team to interview the hotshots, but only as a group. Even the names of the hotshots who spoke during the interview were redacted from the transcripts shared with investigators. The U.S. Forest Service refused to allow the hotshots, or any of its employees, to be interviewed at all for the ADOSH investigation. Documents provided by the Forest Service had so much detail blacked out that investigators complained they were "useless."[9]

The Forest Service's reticence can be traced back to the 2002 Cantwell-Hastings bill, passed as Public Law 107-203, which requires the inspector general of the U.S. Department of Agriculture to investigate burnovers that kill Forest Service employees.[10] That law led to 11 felony charges being filed against the crew boss of four firefighters who perished in a Washington State wildfire. Although the boss was found guilty of only two counts of making false statements, the prosecution of the firefighter made many others, and the organizations they work for, gun-shy about cooperating with investigators.

Investigators found no radio calls from the Granite Mountain crew for 33 minutes before their cry for help and presented the hotshots' last half hour alive as an unknowable mystery and the gap in communications while the crew was moving as a blunder on the part of its leadership. But as soon as the investigations were completed, the hotshots seemed to speak from the grave, as they would for more than a year. Their first shouts were channeled by Holly Neill, a former wildland firefighter who lives in New Mexico. She listened to hundreds of hours of radio traffic after the tragedy and found a few muffled calls from Eric Marsh in the background of recorded radio conversations from an air tanker that was being filmed for Forest Service research. The calls are extremely difficult to understand and there is great disagreement about what Marsh says and uncertainty about what his words may mean, but Neill and some others believe he reported that he was "at the house," presumably indicating he was in the vicinity of the Boulder Springs Ranch safety zone.[11] But then how was he back with the crew when they were overrun?

IN THE YEAR AFTER THE TRAGEDY, the city, fire department, and family members debated whether to rebuild the Granite Mountain Hotshots. Amanda Marsh argued her way to the microphone at a city council meeting to plead for the city to rebuild the hotshots as the best way to honor their legacy. Others, including many wildland firefighters, contended that creating another crew, after all but one member of the original crew had been killed, would just put more firefighters' lives at risk. Some worried that having a wildland crew as part of a department of structural firefighters couldn't help but bias the forest firefighters toward protecting structures — something that they are neither equipped nor trained to do, and that ADOSH had cited as a leading cause of the tragedy. It's reasonably easy to walk away from a bunch of burning trees when things get too dangerous, but much more difficult to step back from burning houses, particularly when they are in a community where the firefighters might know the residents. And with many members of the Granite Mountain crew moving on to the structure firefighting side of the department, it is easy to see how some of them might have wanted to prove their ability to save homes while working in the Wildland Division.

In the end, the decision about whether to rebuild the crew was taken out of the city's hands. Prescott's insurance pool announced that it would drop the city if it had a hotshot crew.[12] A few months later the Northwest Fire District, outside Tucson, Arizona, disbanded its Ironwood Hotshots.[13] The district denied that the destruction of the Prescott crew was a factor in its decision, citing instead the expense, despite the fact that, like the Granite Mountain Hotshots, they were nearly cost neutral.

"I DIDN'T NEED TO SEE THAT," Wade Ward, a former member of Prescott's wildland team who had moved to the structure firefighting side, told me as we drove into the Cathedral Pines suburb of the city exactly a year after the Yarnell Hill Fire had exploded. Ward was the spokesman for the fire department during the tragedy, until city politics and the stress of constantly having to talk about his fallen colleagues drove him back to working on a fire truck and eventually to leave the fire department altogether.

Ponderosa pines and chaparral grew thick among many of the large homes. As we crested a hill, we came upon a group of men in fire department T-shirts dragging brush and tree limbs to the curb. Next to them was the Granite Mountain supervisor's truck, which had been reassigned to the fuels crew — all that remained of Prescott's wildland firefighting operation. Even it was looking at cuts.

"Those guys don't even know what's coming," Ward said when he saw the truck once driven by his boss and friend Eric Marsh. "Most of them are about to lose their jobs."

The one-year anniversary of the tragedy, rather than bringing Prescott, its fire department, the families of the fallen hotshots, and the town they died protecting closer together, just seemed to highlight the rifts growing between them. Hundreds gathered in front of the Yavapai County Courthouse for the city's official ceremony, but the firefighters' families held their own private service at the hotshots' graves in the small memorial park built for them in the Arizona Pioneers' Home cemetery. The town of Yarnell held its own memorial ceremony in a small park along U.S. 89. The large placards that had been set up near the Ranch House Restaurant immediately after the disaster had been dismantled. That early memorial had included large photos of the hotshots; dozens of flags, flowers, toys, and other photos; and a pair of binoculars that allowed visitors to see the canyon where the crew had perished. New displays in the town's park had no view of the site and were bare but for a few plastic flowers and a firefighter's boot.

A coincidence of legal and government timing heightened the tensions. Family members and homeowners who intended to sue had one year to initiate their legal actions, and the hotshots had died on the last day of the fiscal year. So the hours leading up to the one-year anniversary of the tragedy were filled with a cascade of lawsuits, along with the announcement of how the city would fund the fire department going forward. Not only would the Prescott Fire Department not rebuild the hotshot crew, but it also would shutter the Wildland Division that just one year earlier had been a model for the nation. Even the city's structure firefighting operation would feel the pain. By the end of 2015, "rolling brownouts" would close city fire stations for a day at a time to save money.

Darrell Willis watched as the city dismantled most of the 20 years of progress he had made in protecting Prescott from wildfire. He retired in March 2015, but not before accusing the city of tampering with the report used to justify getting rid of the Wildland Division by removing parts of the report that warned of the potential for hundreds of millions of dollars in damage to Prescott from a wildfire.

"I have continued with my commitment to do everything in my power to protect the citizens of Prescott from their greatest danger or risk: Wildfire," Willis wrote in a letter explaining his decision to resign. "It seems the handwriting is on the wall, and the city wants to take a greater risk than any year previous by doing away with the Division that mitigates this great risk of catastrophic wildfire.

". . . Certain people within leadership would even go so far to do away with the Division that they would stoop to the point of tampering with the [International City/County Management Association] Report on the Wildland Division."[14]

That, however, was not his most incendiary claim.

IN OCTOBER 2014, more than 15 months after the Yarnell Hill Fire, Brendan McDonough called Willis. It would be six months before the rest of Prescott heard a version of what the lone survivor of the Granite Mountain Hotshots shared with the man who had created the crew, and that story would be twisted by the tensions between the municipal government and its firefighters when the city attorney, Jon Paladini, put it on the record.

With the potential of Donut's revelations to upset both the wrongful death lawsuits and the workplace safety fine against the state forestry department, Willis said he couldn't sit on them. Willis gave Donut the weekend to consider how he would share the information, after which

Willis told what he had learned to the city attorney and the state forester.

According to Paladini, Donut told Willis that while he was moving the Granite Mountain buggies after he retreated from his lookout post, he heard communications between Eric Marsh and Jesse Steed. Marsh had made it to the safety zone of the ranch and told his second-in-command to bring the rest of the crew there.

"My understanding of the argument between Eric Marsh and Jesse Steed . . . was that Steed did not want to go down," Paladini told the *Arizona Republic*.[15]

The attorney reported that members of the Blue Ridge Hotshots, the federal crew that wouldn't talk with state investigators, also heard the radio traffic. They continued to decline comment after Donut's revelations.

Steed followed the order, Paladini claimed, but protested that it was a very dangerous plan. Marsh headed back into the canyon to help his crew, but by then they were probably already in deep trouble.

"We're not going to make it," Steed radioed to Marsh, according to the attorney.

"Yeah, I know," Marsh allegedly responded. "I'm sorry."[16]

Both Willis and Donut vehemently disputed how Paladini presented the conversation. Willis told me that Donut had recounted a discussion, not an argument, and that Paladini may have been taking vengeance on the wildland chief for supporting the seasonal hotshots in their fight for city benefits. After the city attorney's disclosure, Donut wouldn't discuss what he had heard again, even with Willis. Nonetheless, the disclosure set off a cascade of reports to the Arizona State Forestry Division, the Prescott City Council, and the Arizona Attorney General's Office. Donut received a subpoena to testify but retained a lawyer and didn't appear at three scheduled depositions. Post Traumatic Stress Disorder, he said, would make it difficult to talk with the attorneys about his fallen friends.

ON JUNE 29, 2015, the day before the second anniversary of the tragedy, 12 families of the hotshots and the Arizona State Forestry Division announced a settlement. Each of the families who had sued the state for millions would receive only $50,000. More important to them, they said, was a list of 32 changes the forestry division would make, or had already made, in the way Arizona fights wildfires, including new training for incident commanders and firefighters, national education about the changing nature of fires in drying and warming landscapes, and research into devices to track firefighters.

The forestry division, in lieu of the $559,000 fine assessed by ADOSH, would pay each of the seven families not involved in the lawsuit $10,000 each.[17]

At the press conference in Phoenix announcing the settlement, Roxanne Warneke, the wife of hotshot William Warneke, held their daughter, Billie Grace, who had been born six months after her father's death. She and the mother and wife of deceased hotshot Andrew Ashcraft announced that they would donate their portions of the settlement to a new nonprofit they were forming to promote wildland firefighter safety and fund independent investigations into firefighting accidents. (Amanda Marsh also created a nonprofit to serve wildland firefighters.) Warneke recalled visiting the site where the hotshots had deployed their shelters two weeks after the tragedy and knowing that her husband, a former marine, would never have gone there unless he was ordered to.

"As I stood at that flagpole, I was able to see charred cacti, blackened boulders, and blackened dirt. But what took my breath away was the topography," Warneke said. "I had remembered a saying that my husband had once told me when we were deer scouting two years before. He said to never go into a canyon. That inside the canyon would be thick brush, and how wind travels through a canyon. I know my husband's military and firefighter training. Descending into that canyon went against everything that he had ever been taught . . . I was enraged.

"It prompted me and my family to file a suit against the State of Arizona to prevent another tragedy like this from happening. To prevent Yarnell Hill. To prevent the Dude Fire, the South Canyon Fire, the Thirtymile Fire, the Esperanza Fire, and the Cramer Fire from ever happening again."

Why had her husband descended into that death trap?

"Orders," she said. "He was a marine who was used to taking orders."[18]

Firefighters note that wildland fire crews don't operate like the marines, and every member of a crew is encouraged to speak out about safety issues and decline assignments they believe are too dangerous. Yet many in the firefighting community also were enraged, in their case that a $220 million lawsuit was settled for $670,000 and "good faith" commitments to improve some of the situations that brought on the disaster. With the primary question of what had led to the deaths of the Granite Mountain Hotshots — why they had moved from the black into the peril of the canyon — still largely unanswered, the settlement may have brought closure to the families, but it also may have closed the door on the possibility that there would be "lessons learned" to prevent similar firefighting disasters in the future.

By the time of the second anniversary of the disaster, Donut had missed those three depositions in which he would have told attorneys what he had heard and seen in the moments before his crew was killed. At the press conference announcing the settlement Pat McGroder, the attorney representing the 12 families in the lawsuit, expressed uncertainty as to whether Donut would one day tell what he knew. But he reserved his disdain and anger for the U.S. Forest Service, which forbade its employees to talk to the lawyers and investigators.

"The idea that the federal government is withholding information . . . speaks to the lack of understanding and empathy that they should have for these families," he said. "We would publicly call for . . . the national Forest Service to let their people talk."

Ten months after the press conference announcing the settlement, even Donut was asking for the investigation to be reopened.

"I think with the investigation . . . there's definitely some things that have been found since then, since the investigation. I believe it needs to be opened up again," he said on a Phoenix news radio program. "Some . . . certain things need to be added to it because . . . any decision that was made that day led to their death . . . and we need to learn from that . . . and the wildfire community needs to have those answers and those lessons so that we can prevent this again."[19]

• • •

WILDFIRE TRAGEDIES ARE RARELY the result of one person's decisions, and that was certainly the case in Yarnell. Yet much of the scrutiny of that disaster has focused on one man who isn't alive to explain or defend his actions. Whatever misjudgment Eric Marsh may or may not have made, he also exhibited a heroism that, to me, seemed to bookend the myth of "Big Ed" Pulaski from a century before. Pulaski is legendary for sticking with his crew through the Big Blowup, leading them to a mine shaft, and saving most of their lives by holding them at gunpoint inside it. Marsh, at least some evidence indicates, had parted from his crew, and he would have survived had he stayed at the ranch. Instead, somehow, he rejoined his men as the fire engulfed them.

But whereas Pulaski's crew was retreating from a burning forest, one of the few explanations for Marsh's crew exposing themselves to such avoidable peril was that they were trying to head off an inferno before it destroyed a town. If no homes had been threatened, it's hard, at least for me, to imagine the hotshots would have taken the path that led to their deaths.

But while they saved little in Yarnell, Marsh and his crew made a big difference in other forests and towns across the country. In addition to the hundreds of homes his crew made safe in Prescott with their efforts to reduce the fuel load in the city, his dogged work to create the Arizona Wildfire Academy, and the lives he changed as leader of the Granite Mountain Hotshots, I found another gift Eric Marsh gave to communities confronting fire.

The seed of mythology planted a century ago in the wreckage of the Big Blowup grew into a deeply rooted tree with myriad legends branching off it. One was that we could eradicate natural fire from forests and fields as if it were an unwanted pest. Another, written in the sky over the flames, convinced the public and politicians that planes and retardant could contain every forest fire. Yet another was that new technologies—better fire shelters, more powerful computers—could allow men and women to stand up to conflagrations that were growing larger, faster, and hotter. Then there was the dogma that loggers and grazing animals could take the place of flames in maintaining the forest. And finally there was the delusion that we could build our homes ever deeper

into the nation's most flammable landscapes without facing any consequences.

If the deaths of Eric Marsh and the Granite Mountain Hotshots
trimmed a few branches from that tangled tree of legends, it could save
lives and homes in the future. Cutting down that towering tree altogether, however, will require America to see past the fantasies inspired
by the dancing flames.

EPILOGUE

A YEAR AFTER THEY BURNED, most of the homes destroyed in
the Waldo Canyon and Black Forest Fires in Colorado Springs and the
High Park Fire above Fort Collins had already been rebuilt or were under construction. But in Jefferson County, where the controlled burn
set by Colorado state foresters blew up into the Lower North Fork Fire
that destroyed 23 homes and killed Ann Appel and Sam and Moaneti
Lucas, the burnt Kuehster Road neighborhood was still largely barren.
Two years after the fire, only five homes had been rebuilt.

The survivors, however, felt like they had been repeatedly burned. A
total of 132 claims had been filed against the state, but Colorado law allowed for only $600,000 total in compensation. That led to a two-year
battle between the Colorado attorney general, who fought any settlement above that amount, and the families, several judges and legislators,
and the state claims board. Nearly 30 months after the fire, surviving
families began receiving their portion of the $18.1 million the Colorado
legislature finally approved to compensate them.[1]

Dave Brutout, the firefighter who took it upon himself to warn the
residents of the Kuehster Road neighborhood, gave up fighting fires after more than 20 years. Rather than being honored for getting people
out of the path of the blaze, he was crushed by grief and lawsuits over
the lives and property he couldn't save.

Despite the chaos the under-resourced Elk Creek Fire Department
experienced in the blaze, Chief Bill McLaughlin, who had hoped to update his department's radios and maps even before the fire, seemed
forced to take one step back for every step he took to improve the de-

partment. Ongoing economic difficulties cut his budget by 12 percent the year after the blaze and, while he received a substantial budget increase the year after that, local and state politics presented his department with almost annual financial uncertainty.

COLORADO, HOWEVER, took one of the most expensive options to protect against future blazes.

After the disastrous 2012 wildfire season, several legislators argued that the state's losses from wildfires would have been lower if it had had its own fleet of aircraft to fight them. The following year Steve King, a Republican state senator, pushed for the state to invest in helicopters, large air tankers, and single engine retardant bombers. The legislature approved a $17.5 million bill to create the Colorado Firefighting Air Corps.

Governor John Hickenlooper resisted the idea, largely because of its cost, and the bill provided no way to fund the fleet, leaving the plan effectively dead in the air. Opponents noted that during Colorado's most destructive fires in 2012, when aircraft weren't fighting the fires, it wasn't because they weren't available, but because high winds made aerial attacks unsafe and ineffective. Aerial resources were almost always available from private contractors for considerably less than the Colorado fleet would cost.

King, however, found an unusually persuasive ally. Bill Scott lived outside Colorado Springs when the Waldo Canyon and Black Forest Fires destroyed nearly 1,000 homes there. He worked for the American Center for Democracy, a conservative think tank. He was also a former bureau chief for *Aviation Week and Space Technology* magazine and a member of the blue-ribbon committee convened 10 years earlier to study aerial firefighting after a series of deadly air tanker crashes.

Just before the fires, Scott saw the article in Al Qaeda's online magazine that promoted igniting forest fires as a tool of terrorism. He used the article as the foundation for a congressional briefing in Washington, D.C., titled "Fire Wars," in which he wondered whether America was under seige by terrorists starting forest fires and if the Waldo Canyon Fire that devastated Colorado Springs was one of their successful attacks. He recommended everything from satellites to "fire combat air patrols" to fight the pyro-terrorists. A video of the talk got more than half a million views on YouTube, one of which was by King, who recruited Scott to help push the Colorado Firefighting Air Corps to state legislators in Denver.

"Bill," Scott recalled King telling him, "your job is to scare the hell out of them."

The following year the proposal to create the Colorado Firefighting Air Corps was back, and this time Governor Hickenlooper and the legislature funded a $20 million plan for aerial firefighting. It provided the state with a small plane equipped to locate fires within an hour after they were reported, helicopters, single engine air tankers, and funds for the Center of Excellence for Advanced Technology Aerial Firefighting. The bill didn't budget for large air tankers, of which King wanted four, but he hoped to coax the state to bring these on board in future years.

The high-tech reconnaissance plane, which can see not only a campfire in the forest from thousands of feet in the air but also all of the hunters gathered around it, seemed like a worthy tool to most firefighters, but the value of the state fleet of firefighting aircraft was questioned by many.

"Colorado must have a lot of money . . . to waste," said Andy Stahl, the executive director of Forest Service Employees for Environmental Ethics.

Stahl noted that state officials were in much the same position that leaders of firefights find themselves in when they are pressured to fight fires with aircraft that they know won't make a difference. "All of the incentives are geared to cause you to throw everything in your tool kit at the fire," he told me. "Because there are no cost controls whatsoever."

Ray Rasker at Headwaters Economics, the think tank that studied the cost of the nation's wildfires, also saw Colorado's investment in firefighting aircraft as wasteful, but he was happy to see the costs borne at the state level rather than by federal taxpayers. If more of the tab for wildfire protection was paid by the people who live in flammable landscapes, he said, they would be less likely to build in the path of wildfires.

Even Jerry Williams, former director of fire and aviation for the U.S. Forest Service, found that the problem wasn't that there weren't enough planes, but that the nation's strategy for dealing with wildfires continues to focus on fighting them rather than building homes and communities that are more resilient to fire and letting natural wildfires far from development burn to avoid putting planes and firefighters at risk.

"You can get lost in the weeds... about the efficacy of night air tanker operations, or better retardant mixes, or bigger helicopters, or more engines," he said in an interview. "But if you overlook the basic strategic issues here, all is lost."

THE FEDERAL GOVERNMENT TRIED MANY TACTICS to stem its hemorrhaging of money to fight wildfires. In March 2014 the Obama administration proposed changing how the government funds fighting that 1 percent of fires that, according to the White House, consume more than 30 percent of federal wildfire suppression money.

The president's proposal, and bipartisan bills introduced in the House and Senate, would give the Forest Service and Interior Department access to the Federal Emergency Management Agency's Disaster Relief Fund. Created in 2011, the emergency fund was earmarked for relief efforts in hurricanes, tornadoes, and major floods. The recommended change, which required congressional action, would add large wildfires to the list. FEMA funds would be tapped once the costs of managing wildfires reached 70 percent of their 10-year average, similar to how other disaster recovery efforts are funded.

"The President's budget proposal, and similar bipartisan legislation before Congress, would solve a recurring problem of having to transfer money from forest restoration and other Forest Service accounts to pay the costs of fighting wildfires," said Secretary of Agriculture Tom Vilsack in a press release discussing the provision. "USDA will spend the necessary resources to protect people, homes and our forests, but it is not in the interest of forest health to transfer funds from forest restoration that can prevent future fires."[2]

Proponents argued that the $12 billion Disaster Relief Fund, which falls outside discretionary budget limits, had gone largely unspent in recent years, giving the Obama administration a clear solution to addressing wildfire funding shortfalls.

"Fire is every bit as much an emergency as a tornado or a hurricane," said Representative Mike Simpson, a Republican from Idaho, who supported the Obama administration's proposal.

But Republicans east of the Mississippi, and former House Budget Committee chairman Paul Ryan in particular, resisted giving federal firefighters access to the emergency funds.

Andy Stahl, who saw wasteful spending as the biggest reason for the rise in federal firefighting costs, called the administration's proposal "another blank check for firefighting that does nothing to contain costs."

Despite the fact that nearly 150 members of Congress supported the bill, it never came up for a vote in 2014.

Five days into the next legislative session, representatives from both parties reintroduced the bill. In the following months a record number of U.S. acres burned, and another firefighting disaster occurred at the Twisp River Fire in Washington State, where the second exploding wildfire in two years trapped four wildland firefighters in their truck, killing three of them. Congress again failed to vote on the bill.[3]

WITH THE WEST CERTAIN to have much more fire, we'll increasingly be faced with choices about what kind of relationship we have with it, rather than whether we allow it to burn at all.

Two years after the Colorado State Forest Service's prescribed burn turned into the deadly Lower North Fork Fire in Jefferson County, Colorado, I stood with Jay Stalnacker, the fire management officer with the Boulder County Sheriff's Office, at Heil Valley Ranch, about five miles north of Boulder. Five thousand acres of hanging valleys and ponderosa pine forests make up the ranch — the largest of the county's open spaces and a haven for wildlife — but grazing and fire suppression had left many of its forests overgrown.

I'd visited the ranch with University of Colorado professor Tom Veblen, who'd pointed to it as an example of the portion of Front Range forests that exhibit an increase in wildfires due to previous fire suppression and grazing causing the forest to grow unnaturally dense. A week later I'd run into Rod Moraga, the fire behavior analyst who'd lost his home in the Fourmile Canyon Fire, as he mountain biked through the ranch to take photos that would help in planning a prescribed burn there to bring the forest closer to its historic density of trees and ground cover.

In October 2014 I joined Moraga, Stalnacker, and about 50 other firefighters on the first morning of the burn. After the Lower North Fork disaster in March 2012, Governor John Hickenlooper had banned all prescribed burns for nearly a year, and when they were allowed to resume, new laws dictated how they were to be managed.

The leaders at Heil Valley Ranch carefully went over their preparations for the fire. Eight acres of blackline and portable water tanks surrounded the area they planned to burn. Local fire departments and a helicopter were on call. A public information officer contacted nearby homeowners and the media. Moraga provided details of the weather forecasts for the day of the burn and the coming week.

The crew hoped to create a fire that would kill about 40 percent of the mature pines, leaving about 60 trees per acre, and reduce the slash on the ground by about one-half.

Stalnacker, a former smokejumper, was the burn boss. As he briefed the crews, it was clear that he saw something more important than just making one forest healthier. He brought up the topic nobody wanted to talk about — the deadly prescribed burn two years earlier.

"One of the biggest things we lost that day was public trust," he said. "You have a chance today to regain that public trust. I ask that you connect with this piece of land."

Then he asked them to recognize their place in history.

"We were there when we put fire back in the Front Range."

A few hours later the firing crew dripped flames from their torches onto the forest floor. The flames spread lazily along the ground, rarely rising more than a foot into the air. On occasion a tree inside the burn zone torched dramatically. Firefighters surrounded the area and walked carefully through it, occasionally stepping over the flames.

Months later I biked through and saw that most of the charred ground already had new grass. Some torched pines had fallen, but most were still standing. Many of those would die to make room for survivors that would bear "cat face" fire scars from the burn when they grew into massive trees over the coming decades. Most of the cyclists and hikers on the trail around me didn't notice the burn at all. But during the following weeks and years I often saw smoke rise from the open space on days when the weather conditions were conducive to holding prescribed burns.

JEN STRUCKMEYER, THE VOLUNTEER FIREFIGHTER who was burned over on the Colorado plains during the winter of 2012, learned to walk again a few months before the Granite Mountain Hotshots perished. She'd spent nearly three months in the hospital and at a rehab center. By then the pain from the amputations to her burnt foot and the scars on her leg and arms had eased a bit. On the wall of her house was a framed purple heart honoring her service at the Heartstrong Fire. On her truck, a sticker read "Fight Like a Girl."

She, however, would never fight fire again.

Instead, after her recovery, she volunteered as an emergency medical technician, first in the town of Holyoke, near the Struckmeyer ranch, then with the City of Yuma Ambulance Service, the same department that rescued her after she was burned over in the Heartstrong Fire. Eventually she would help deliver several burn victims to the hospital herself. The first time, when the victim's family prayed over him, Jen remembered how her own family had gathered around her hospital bed to pray. The rush of emotions drove her out of the room, but once outside, she stopped and calmed down.

"You can do this," she told herself. She stepped back into the room to tell the victim what he could expect, and that he was strong enough to survive his burns.

Jen's husband, Del — the burliest of the three Struckmeyer brothers on the Wages Volunteer Fire Department — tried to remain loyal to the service his family had built. While his wife was still recovering, a fire call came in, and he responded with his sister-in-law Pam, who had been in the truck when the family was overrun in Heartstrong. But as they approached the glow of the flames, Del felt something he'd never encountered in 20 years of firefighting — panic.

"I can't do this anymore," he told Pam.

They turned around, and Del drove back to his house, where he sat silently with his wife.

ACKNOWLEDGMENTS

This book started with a simple idea and turned into an epic journey, during which I was dependent on the help of hundreds of people. First and foremost, I'm deeply indebted to the wildland firefighters, their family members, and survivors of wildfires around the world who shared their time and stories with me. In Prescott, Arizona, David and Claire Caldwell, Danny Parker, Wade Ward, Darrell Willis, Conrad Jackson, Brendan McDonough, and Pat McCarty were particularly helpful, as were many of the staff and participants at the Arizona Wildfire and Incident Management Academy.

In Colorado the Struckmeyer family welcomed me into their homes, and Damon, Del, Jen, and Pam Struckmeyer, along with Darin Stuart, generously relived a horrifying experience from which they were still recovering. Sheriff Chad Day and Chief Elmer Smith made sure I understood the challenges faced by the volunteers at the Heartstrong Fire. At the Lower North Fork Fire, Bill McLaughlin, Andy and Jeanie Hoover, Kim Olson, Tom and Sharon Scanlan, Kristen Moeller and Dave Cottrell, and Dave Brutout shared their experiences while still digging out from the tragedy and its aftermath. Steve Riker, Jim Schanel, Dave Vitwar, Steve Wilch, and Steve Schopper from the Colorado Springs Fire Department recounted their efforts during the Waldo Canyon and Black Forest Fires, while Cindy Maluschka, Michael and Peri Duncan, and Laura Hunt helped me understand the impacts of those wildfires on suburban residents.

Rod Moraga was particularly helpful not only in my understanding of the Fourmile Canyon Fire that took his home, but also in all aspects of Colorado's and the nation's response to fire disasters, and in the challenges of mitigating developments in dangerously flammable forests. That we could do some of that while on bikes and skis helped make my research more tolerable.

The chief of the U.S. Forest Service, Tom Tidwell, and the service's now retired directors of fire and aviation management Jerry Williams and Tom Harbour, along with Randy Eardley, Jennifer Jones, Susie Stingley-Russell, Tim Murphy, Scott Fisher, John Segar, Ed Delgado, and Jeremy Sullens at the National Interagency Fire Center, guided me through the nation's complex systems for predicting, preparing for, and responding to wildfires throughout the nation. My own fire crew in Connecticut years ago tolerated my photography and note taking and made sure those never got in the way of my safety or that of my colleagues, a skill that continues to serve me today.

I've spoken to scores of researchers focused on the earth's fire cycles and history. Most helpful among them were Tom Veblen and the crew in the terrific Biogeography Lab he leads at the University of Colorado. Stephen J. Pyne, Craig Allen, Jack Cohen, Mark Finney, Tom Swetnam, Tania Schoennagel, Jennifer Balch, Max Moritz, Ray Rasker, Russ Braddock, David Bowman, LeRoy Westerling, Richard Wrangham, Chad Hanson, George Wuerthner, Park Williams, Mike Battaglia, Dave Theobald, Dave Lucas, Steve Segin, Chad Oliver, Sergiy Zibtsev, Gavriil Xanthopoulos, and the staff at the Tall Timbers Research Station helped me understand the many facets of wildfire science, economics, and management. Nolan Doesken, Katharine Hayhoe, and Jim White explained the complex relationship between climate and fire. Jeremy Bailey, of the Nature Conservancy, and Jay Stalnacker, with the Boulder County Sheriff's Office, welcomed me to the prescribed burns they held in Nebraska and Colorado and guided me through the challenges of reintroducing fire to our forests. Bob Mutch, Steve Arno, and George Weldon provided excellent insights into the U.S. Forest Service and wilderness fires, as well encouragement for my work.

Tim Rasmussen, Helen Richardson, Mahala Gaylord, AAron Ontiveroz, R. J. Sangosti, and Bruce Finley at the *Denver Post* helped me develop this project, along with several smaller ones, and Laura Frank and

Burt Hubbard at Rocky Mountain PBS supported and collaborated on my wildfire work from the beginning. John Maclean, a reporter and author I've long admired, provided guidance early on in this project. Holly Neill read and refined critical portions of the manuscript.

I owe a special debt to Tom Yulsman, Len Ackland, and Cindy Scripps for selecting me, in 2009, as a Ted Scripps Fellow in Environmental Journalism at the University of Colorado's Center for Environmental Journalism, the nursery where the seed of this project was first planted. A few years later, when the CEJ turned into my academic and professional home as a faculty member in the University of Colorado's new College of Media, Communication and Information, I found myself with a newsroom of journalism fellows and students, many of whom deserve deep gratitude. First, I'm thankful to Chris Braider, who hired me. Tom Yulsman continued to mentor me as closely as a colleague on the faculty as he had as a codirector of my fellowship.

Several of my graduate students were critical in the completion of this project. Gloria Dickie, in particular, provided crucial reporting on wildfire and invasive species, particularly cheatgrass, and collaborated with me on research and reporting in Colorado, Arizona, and Montana. She assisted with some of the most emotionally demanding reporting in Prescott, took the lead in reporting on the Last Chance Fire, and provided a valuable read of the book during one of its many transformations. Christi Turner reported on the Fern Lake Fire and also provided reporting on wildfire and climate, and on black carbon's impacts on snow and glacial ice. Caitlin Rockett helped report on the Arizona Wildfire Academy and the history of Eric Marsh and the Granite Mountain Hotshots in the development of that training facility. Paul McDivitt helped report on the nation's wildfire budget challenges and did much of the reporting on Chernobyl's risk of nuclear wildfires. Kelsey Ray dug deep into the creation of Colorado's fleet of firefighting aircraft and fire aviation center. She also assisted with the reporting on the Waldo Canyon Fire in Colorado Springs and gave a valuable read of the early manuscript. Avery McGaha reported on "vapor-pressure deficit" and how that is stressing southwestern forests. Every student who has taken my Reporting on the Environment class deserves my gratitude for tolerating my obsession and assignments related to wildfire. Kevin Moloney, who taught the Transmedia Wildfire class alongside me, deserves special thanks for his insights, experience, unflappable demeanor, and terrific teaching abilities.

Hillary Rosner provided great support and an early read of my manuscript. Frank Allen and the staff at the Institute for Journalism and Natural Resources provided me access to the forest charred by the Las Conchas Fire and important sources who continue to guide me. Many friends in the Society of Environmental Journalists have provided guidance and support for my interest in wildfires and other environmental hazards over the years, and I continue to be amazed by what a small, scrappy, underfunded organization can accomplish with the right combination of passion, smarts, and commitment. My agent, Wendy Strothman, was a valuable guide throughout the unexpected twists, turns, and time sucks involved in creating this book, and my editor, Susan Canavan, showed patience and good humor as the subject grew into something far larger than any of us ever anticipated. Copyeditor Barbara Jatkola's diligence and precision saved my manuscript more times than I care to mention.

My brother Jeff Polson and his partner, Nara Wood, generously recounted their experiences in the Valley Fire, cheerfully recounting their losses and grief. My other brothers, Todd Kodas and Doug Polson, put up with my fascination with wildfire and the long, tedious process of turning that into a book.

My wife, Carolyn Moreau, tolerated many "vacations" to fire disasters around the world and managed a variety of steering wheels for hours on end while I tapped on my laptop in the passenger seat.

Finally, I'd like to thank my mother, Anita, who didn't see this project come to its fruition but planted the seed that grew into it.

NOTES

PROLOGUE

1. CAL FIRE Incident Management Team 3, Valley Incident Damage Inspection Team, "Valley Incident Damage Inspection Report" (September 12, 2015), http://cdfdata.fire.ca.gov/pub/cdf/images/incidentfile1226_1957.pdf.
2. U.S. Department of Agriculture, "Agriculture Secretary Tom Vilsack Announces 2015 Wildfires Burned Record Acres, Urges Congress to Pass Wildfire Funding Fix," news release, January 6, 2016.
3. Jerry Williams et al., "Findings and Implications from a Course-Scale Global Assessment of Recent Selected Mega-Fires" (paper presented at the Fifth International Wildland Fire Conference, Sun City, South Africa, May 9–11, 2011).

1. YARNELL HILL

1. Jim Karels et al., "Yarnell Hill Fire, June 30, 2013: Serious Accident Investigation Report" (State of Arizona, September 23, 2013), http://www.iawfonline.org/Yarnell_Hill_Fire_report.pdf.
2. Ibid.
3. Ibid.
4. Ibid.
5. Charles L. Myers et al., "Coal Canyon Fire, August 11, 2011: Serious Accident Investigation Report" (Hell Canyon Ranger District, Black Hills National Forest, South Dakota, 2012), https://www.fs.fed.us/rm/pubs_other/rmrs_2012_meyers_c001.pdf.
6. "Yarnell Hill Fire Fallen Remembered: Wade Parker," *Arizona Republic,* July 10, 2013.
7. Karels et al., "Yarnell Hill Fire."

2. FUSES AND BOMBS

1. David E. Calkin, Matthew P. Thompson, and Mark A. Finney, "Negative Conse-

quences of Positive Feedbacks in US Wildfire Management," *Forest Ecosystems* 2, no. 9 (December 2015).

2. Carter Stone, Andrew Hudak, and Penelope Morgan, "Forest Harvest Can Increase Subsequent Forest Fire Severity," in *Proceedings of the Second International Symposium on Fire Economics, Planning, and Policy: A Global View* (U.S. Forest Service, Pacific Southwest Research Station, General Technical Report PSW-GTR-208, April 2008).

3. "Fire Outlook," U.S. Department of Agriculture, https://www.usda.gov/topics /climate-solutions.

4. Sebastián Martinuzzi et al., "The 2010 Wildland-Urban Interface of the Conterminous United States," U.S. Forest Service, 2015.

5. David M. Theobald and William H. Romme, "Expansion of the US Wildland-Urban Interface," *Landscape and Urban Planning* 83, no. 4 (December 2007): 340.

6. Headwaters Economics, "The Wildland-Urban Interface: Trends, Future and Solutions" (June 3, 2016), https://headwaterseconomics.org/wphw/wp-con tent/uploads/wildfire_homes_solutions_presentation.pdf.

7. U.S. Forest Service, "The Rising Cost of Wildfire Operations: Effects on the Forest Service's Non-Fire Work" (August 4, 2015), https://www.fs.fed.us/sites/de fault/files/2015-Fire-Budget-Report.pdf.

8. Ross Gorte, "The Rising Cost of Wildfire Protection" (Headwaters Economics, June 2013), https://headwaterseconomics.org/wp-content/uploads/fire-costs -background-report.pdf.

9. Andrew Freedman, "The Climate Context Behind the Deadly Arizona Wildfire," Climate Central, July 1, 2013, http://www.climatecentral.org/news/the -climate-context-behind-the-deadly-arizona-wildfire-16175.

10. Peter Z. Fulé, W. Wallace Covington, and Margaret M. Moore, "Determining Reference Conditions for Ecosystem Management of Southwestern Ponderosa Pine Forests," *Ecological Applications* 7, no. 3 (1997): 895.

11. Mary Jo Pitzl, Brandon Loomis, and Matthew Dempsey, "In Harm's Way," Arizona Republic, December 8, 2013, http://archive.azcentral.com/news/wild fires/yarnell/arizona-wildfires-homes-forests-risk/.

12. 2009 Victorian Bushfires Royal Commission, "The Fires and the Fire-Related Deaths: Final Report," vol. 1 (July 2010), http://www.royalcommission.vic.gov .au/Finaldocuments/volume-1/HR/VBRC_Vol1_Introduction_HR.pdf.

13. Alon Tal, *All the Trees of the Forest: Israel's Woodlands from the Bible to the Present* (New Haven, CT: Yale University Press, 2013).

14. George Monbiot, "Indonesia Is Burning. So Why Is the World Looking Away?," *Guardian,* October 30, 2015, https://www.theguardian.com/comment isfree/2015/oct/30/indonesia-fires-disaster-21st-century-world-media.

15. Kate Lamb, "Indonesia's Fires Labelled a 'Crime Against Humanity' as 500,000 Suffer," *Guardian,* October 26, 2015, https://www.theguardian.com /world/2015/oct/26/indonesias-fires-crime-against-humanity-hundreds-of -thousands-suffer.

16. "Indonesia's Fire and Haze Crisis," World Bank, November 25, 2015, http://www.worldbank.org/en/news/feature/2015/12/01/indonesias-fire-and-haze-crisis.

17. Nancy Harris et al., "Indonesia's Fire Outbreaks Producing More Daily Emissions Than Entire US Economy," World Resources Institute, October 16, 2015, http://www.wri.org/blog/2015/10/indonesia%E2%80%99s-fire-outbreaks-producing-more-daily-emissions-entire-us-economy.

18. Eric Holthaus, "Wildfire Rips Through Canadian City, Forcing 80,000 to Flee. This is Climate Change," *Slate*, May 4, 2016, http://www.slate.com/blogs/the_slatest/2016/05/04/fort_mcmurray_alberta_wildfire_forces_major_evacuation.html.

19. Jeff Spross, "Historic Wildfires Burn Through Canada as Sub-Arctic Forests Heat Up," ThinkProgress, August 25, 2014, https://thinkprogress.org/historic-wildfires-burn-through-canada-as-sub-arctic-forests-heat-up-d0b9e186219e#.rwyna73vp.

20. Renee Cho, "The Damaging Effects of Black Carbon," *State of the Planet*, Earth Institute, Columbia University, March 22, 2016, http://blogs.ei.columbia.edu/2016/03/22/the-damaging-effects-of-black-carbon/.

21. "Record Temperatures and Wildfires in Eastern Russia," NASA, August 8, 2012, https://www.nasa.gov/mission_pages/fires/main/world/20120828-russia.html.

22. I. B. Konovalov et al., "Atmospheric Impacts of the 2010 Russian Wildfires," *Atmospheric Chemistry and Physics* 11 (October 4, 2011).

23. Spyros Skouras and Nicos Christodoulakis, "Electoral Misgovernance Cycles" (Hellenic Observatory, European Institute, May 2011).

24. A. L. Westerling et al., "Warming and Earlier Spring Increase Western U.S. Forest Wildfire Activity," *Science* 313, no. 5789 (2006).

25. Anthony LeRoy Westerling, "Increasing Western US Forest Wildfire Activity: Sensitivity to Changes in the Timing of Spring," *Philosophical Transactions of the Royal Society B* 371, no. 1696 (June 5, 2016), doi:10.1098/rstb.2015.0178.

26. "Quadrennial Fire Review 2009" (final report, U.S. Departments of the Interior and Agriculture and National Association of State Foresters, January 2009), https://www.forestsandrangelands.gov/strategy/documents/foundational/qfr2009final.pdf.

27. "Total Wildland Fires and Acres (1960–2015)," National Interagency Fire Center, https://www.nifc.gov/fireInfo/fireInfo_stats_totalFires.html.

28. U.S. Forest Service, "The Rising Cost of Wildfire Operations."

4. HEARTSTRONG

1. "Colorado Fire Loss/Fire Department Profile," U.S. Fire Administration, https://www.usfa.fema.gov/data/statistics/states/colorado.html.

2. Margaret Herzog, "Managing Drought Before It's Time" (American Water Resources Association, Colorado Chapter, April 21, 2012).

3. Wendy Ryan and Nolan Doesken, "Drought of 2012 in Colorado" (Colorado Climate Center, Department of Atmospheric Science, Colorado State University, n.d.), http://ccc.atmos.colostate.edu/pdfs/climo_rpt_13_1.pdf.

4. Scott Irwin and Darrel Good, "How Bad Was the 2012 Corn and Soybean Growing Season?," Farmdocdaily, Department of Agriculture and Consumer Economics, University of Illinois at Urbana-Champaign, October 3, 2012, http://farmdocdaily.illinois.edu/2012/10/how-bad-was-the-2012-corn-and-.html.

5. Yuma County Sheriff's Office, "Incident Report #12078 (Heartstrong Fire)," Yuma, AZ.

6. Ibid.

5. RED BUFFALO, BLACK DRAGON

1. Julie Courtwright, *Prairie Fire: A Great Plains History* (Lawrence: University Press of Kansas, 2011).

2. "Last Stand of the Tallgrass Prairie," Tallgrass Prairie National Preserve, Kansas, https://www.nps.gov/tapr/index.htm.

3. Julie Courtwright, "Taming the Red Buffalo: Prairie Fire on the Great Plains" (PhD diss., University of Arkansas, Fayetteville, 2007).

4. "A Complex Prairie Ecosystem," Tallgrass Prairie National Preserve, Kansas, https://www.nps.gov/tapr/learn/nature/a-complex-prairie-ecosystem.htm.

5. Alan K. Knapp et al., "The Keystone Role of Bison in North American Tallgrass Prairie," *BioScience* 49, no. 1 (1999).

6. Courtwright, *Prairie Fire.*

7. Clenton E. Owensby and John Bruce Wyrill III, "Effects of Range Burning on Kansas Flint Hills Soil," *Journal of Range Management* 26, no. 3 (1973), https://journals.uair.arizona.edu/index.php/jrm/article/viewFile/6183/5793.

8. Sir Charles Augustus Murray, *Travels in North America During the Years 1834, 1835, and 1836* (New York: Harper and Brothers, 1839).

9. "Woman Found Dead in Field Fire in Labette County," *Pittsburg (KS) Morning Sun,* March 15, 2010, http://www.morningsun.net/x313365373/Woman-found-dead-in-field-fire-in-Labette-County.

10. Bob Mutch and Paul Keller, "Lives Lost—Lessons Learned: The '05–'06 TX and OK Wildfires," Wildland Fire Lessons Learned Center, October 2010.

11. Bill Gabbert, "Follow-Up on Oklahoma Fatality," Wildfire Today, January 6, 2009, http://wildfiretoday.com/2009/01/06/follow-up-on-oklahoma-fatality/.

12. Harrison Salisbury, *The New Emperors: China in the Era of Mao and Deng* (New York: Harper Perennial, 1993).

13. Harrison Salisbury, *The Great Black Dragon Fire: A Chinese Inferno* (New York: Little, Brown, 1989).

14. Ibid.

15. Rachel Carmody and Richard Wrangham, "The Energetic Significance of Cooking," *Journal of Human Evolution* 57, no. 4 (2009).

16. John Pickrell, "Human 'Dental Chaos' Linked to Evolution of Cooking," *New Scientist*, February 19, 2005.

17. Courtwright, *Prairie Fire*.

6. CRAZY WOMAN

1. Don Thompson, "California Does Allow Violent Inmates in Its Firefighting Crews After All," Associated Press, *Sacramento Bee*, October 15, 2015, http://www.sacbee.com/news/nation-world/national/article39272541.html.

7. THE BIGGER BLOWUP

1. "1910 Fires: People," U.S. Forest Service, Northern Region, https://www.fs.usda.gov/detail/r1/learning/history-culture/?cid=stelprdb5122868#bettygoodwinspencer.

2. U.S. Forest Service, "A Synopsis of the Pulaski Rescue Story and the Great Fire of 1910" (n.d.), https://www.fs.usda.gov/Internet/FSE_DOCUMENTS/stelprdb5442828.pdf.

3. William James, "The Moral Equivalent of War," *McClure's Magazine*, August 1910, http://www.unz.org/Pub/McClures-1910aug-00463.

4. "U.S. Forest Service Fire Suppression," Forest History Society, http://www.foresthistory.org/ASPNET/Policy/Fire/Suppression/Suppression.aspx.

5. SmokeyBear.com, https://smokeybear.com/en.

8. MANSIONS IN THE SLUMS

1. "Wildland Fire," Disaster Resource Center, U.S. Department of Agriculture, https://www.usda.gov/wps/portal/usda/usdahome?navid=wildland-fire.

2. Boulder County Parks and Open Space, "Boulder County Parks and Open Space Forest Management Policy" (Boulder, CO, May 18, 2010), http://www.colorado.edu/geography/class_homepages/geog_4430_s08/Forest%20Policy.pdf.

3. Peter M. Brown, Merrill R. Kaufmann, and Wayne D. Sheppard, "Long-Term, Landscape Patterns of Past Fire Events in a Montane Ponderosa Pine Forest of Central Colorado," *Landscape Ecology* 14 (1999): 513, doi:10.1023/A:1008137005355.

4. "Bark Beetle F.A.Q.," *Arizona Forest Health*, Arizona Cooperative Extension, University of Arizona, https://cals.arizona.edu/extension/fh/bb_faq.html.

5. A. S. Leopold et al., "Wildlife Management in the National Parks: The Leopold Report," National Park Service, March 4, 1963, https://www.nps.gov/parkhistory/online_books/leopold/leopold.htm.

6. U.S. Department of the Interior and U.S. Department of Agriculture, "Federal Wildland Fire Management Policy and Program Review" (December 18, 1995),

https://www.forestsandrangelands.gov/strategy/documents/foundational/1995_fed_wildland_fire_policy_program_report.pdf.

7. "Healthy Forests Initiative," U.S. Forest Service, https://www.fs.fed.us/projects/hfi/.

8. U.S. Department of Agriculture, "Audit Report: Forest Service Large Fire Suppression Costs" (Office of Inspector General, Western Region, November 2006), https://www.usda.gov/oig/webdocs/08601-44-SF.pdf.

9. "Quadrennial Fire and Fuel Review Report" (U.S. Departments of the Interior and Agriculture, June 30, 2005), https://www.forestsandrangelands.gov/strategy/documents/foundational/qffr_final_report_20050719.pdf.

10. Colorado State Forest Service, "Colorado Statewide Forest Resource Assessment: A Foundation for Strategic Discussion and Implementation of Forest Management in Colorado" (Fort Collins, October 20, 2008), http://static.colostate.edu/client-files/csfs/pdfs/SFRA09_csfs-forestassess-web-bkmrks.pdf.

11. Mark A. Williams and William A. Baker, "Spatially Extensive Reconstructions Show Variable-Severity Fire and Heterogeneous Structure in Historical Western United States Dry Forests," *Global Ecology and Biogeography* 21 (2012): 1042–52, doi:10.1111/j.1466-8238.2011.00750.x

12. Rosemary L. Sherriff et al., "Historical, Observed, and Modeled Wildfire Severity in Montane Forests of the Colorado Front Range," *PLOS One* 9, no. 9 (2014), doi:10.1371/journal.pone.0106971.

13. Richard Hutto, "The Ecological Importance of Severe Wildfires: Some Like It Hot," *Ecological Applications* 18 (2008): 1827–34, doi:10.1890/08-0895.1.

14. "From Forests to Faucets: U.S. Forest Service and Denver Water Watershed Management Partnership," Denver Water, http://www.denverwater.org/SupplyPlanning/WaterSupply/PartnershipUSFS/.

9. THE BLACKLINE

1. Bureau of Land Management, "Memorandum of Interview: Kirk Will," interviewed by Bureau of Land Management Special Agent Shannon Tokos, Colorado State Forest Service, Golden District Office, Golden, CO, March 30, 2012.

2. "Lower North Fork Prescribed Fire: Prescribed Fire Review" (State of Colorado, Office of Executive Director, Department of Natural Resources, Denver; and Office of the President, Colorado State University, Fort Collins, April 13, 2012), http://dnr.state.co.us/SiteCollectionDocuments/Review.pdf.

3. Bureau of Land Management, "Memorandum of Interview: Kevin Michalak," interviewed by Bureau of Land Management Special Agent Shannon Tokos, Lower North Fork Fire, near Prescribed Unit 4, Conifer, CO, March 29, 2012.

4. "Lower North Fork Prescribed Fire."

5. Ibid.

6. Bureau of Land Management, "Memorandum of Interview: Kevin Michalak."

7. Bureau of Land Management, "Memorandum of Interview: Kirk Will."

8. Bureau of Land Management, "Memorandum of Interview: Rocco Snart," inter-

viewed by U.S. Forest Service Special Agent Brenda Shultz, Lower North Fork Fire Incident Command Post, Conifer High School, Conifer, CO, March 30, 2012.

9. "Wildland Fire Investigation," Bureau of Land Management Special Agent Shannon Tokos for the Jefferson County Sheriff's Department, April 12, 2012.
10. "Lower North Fork Prescribed Fire."
11. Wendy Ryan and Nolan Doesken, "Drought of 2012 in Colorado" (Colorado Climate Center, Department of Atmospheric Science, Colorado State University, n.d.), http://ccc.atmos.colostate.edu/pdfs/climo_rpt_13_1.pdf.
12. "Lower North Fork Prescribed Fire."
13. Ibid.

10. SLOP-OVER

1. Bureau of Land Management, "Memorandum of Interview: Kevin Michalak," interviewed by Bureau of Land Management Special Agent Shannon Tokos, Lower North Fork Fire, near Prescribed Unit 4, Conifer, CO, March 29, 2012.
2. Ibid.
3. "Lower North Fork Prescribed Fire: Prescribed Fire Review" (State of Colorado, Office of Executive Director, Department of Natural Resources, Denver; and Office of the President, Colorado State University, Fort Collins, April 13, 2012), http://dnr.state.co.us/SiteCollectionDocuments/Review.pdf.
4. Bureau of Land Management, "Memorandum of Interview: Kevin Michalak."
5. Ibid.
6. "Lower North Fork Prescribed Fire."
7. Bureau of Land Management, "Memorandum of Interview: Kevin Michalak."
8. Bureau of Land Management, "Memorandum of Interview: Kirk Will," interviewed by Bureau of Land Management Special Agent Shannon Tokos, Colorado State Forest Service, Golden District Office, Golden, CO, March 30, 2012.
9. Bureau of Land Management, "Memorandum of Interview: Kevin Michalak."
10. Ibid.
11. "Lower North Fork Prescribed Fire."
12. Bureau of Land Management, "Memorandum of Interview: Kevin Michalak."
13. "Lower North Fork Prescribed Fire."
14. "Wildland Fire Investigation," Bureau of Land Management Special Agent Shannon Tokos for the Jefferson County Sheriff's Department, April 12, 2012.
15. Bureau of Land Management, "Memorandum of Interview: Kevin Michalak."
16. "Lower North Fork Prescribed Fire."
17. Bureau of Land Management, "Memorandum of Interview: Rich Palestro," interviewed by Bureau of Land Management Special Agent Shannon Tokos, Colorado State Forest Service, Golden District Office, Golden, CO, March 30, 2012.

11. OFF TO THE RACES

1. Felisa Cardona, "Colorado Wildfire Victims Sam and Linda Lucas Remem-

bered at Lakewood Service," *Denver Post,* March 30, 2012, http://www
.denverpost.com/2012/03/30/colorado-wildfire-victims-sam-and-linda-lucas
-remembered-at-lakewood-service-2/.

2. Andy Hoover, interview by Ron Zappolo, *Zappolo's People,* Fox 31, KDVR, Denver, http://kdvr.com/2012/05/02/zappolos-people-andy-hoover/.

3. Kirk Mitchell, "Colorado Wildfire Devours Futuristic Engineering, Historical
Treasures," *Denver Post,* April 29, 2012.

4. "911 Told Fire Victim That Fire Was Controlled Burn," TheDenverChannel.com,
April 4, 2012, http://www.thedenverchannel.com/news/911-told-fire-victim
-that-fire-was-controlled-burn.

5. Colleen Curry, "Colorado Wildfire Deaths Blamed on 911 Malfunction," ABC
News, April 4, 2012, http://abcnews.go.com/US/colorado-wildfire-deaths
-blamed-911-malfunction/story?id=16072575.

6. John Ingold and Kirk Mitchell, "Lower North Fork Fire Ambushed Crews, Surprised Residents," *Denver Post,* April 28, 2012.

7. "Raw Video: Escaping the Lower North Fork Fire," DPTV, *Denver Post,* March
28, 2012, YouTube, https://www.youtube.com/watch?v=KdBaqfiynLE.

8. "Deputy David Bruening Narrative, Jefferson County Sheriff's Office," Report
12-8273, Supplement 1, April 2, 2013.

9. Ibid.

10. "Deputy Randal Barnes Narrative, Jefferson County Sheriff's Office," Report
12-8273, Supplement 3, April 2, 2013.

12. RED ZONES

1. Howard Botts et al., "Wildfire Hazard Risk Report: Residential Wildfire Exposure Estimates for the Western United States" (CoreLogic, 2015), http://www
.corelogic.com/research/wildfire-risk-report/2015-wildfire-hazard-risk
-report.pdf.

2. S. M. Stein et al., "Wildfire, Wildlands, and People: Understanding and Preparing for Wildfire in the Wildland-Urban Interface" (U.S. Forest Service, Rocky
Mountain Research Station, General Technical Report RMRS-GTR-99, January 2013), https://www.fs.fed.us/openspace/fote/reports/GTR-299.pdf.

3. "The Blue Ribbon Panel Report on Wildland Urban Interface Fire" (International Code Council, April 4, 2008), https://inawf.memberclicks.net/assets
/blueribbonreport-low.pdf.

4. Headwaters Economics, "Montana Wildfire Cost Study Technical Report"
(August 8, 2008), http://www.headwaterseconomics.org/wildfire/Headwaters
Economics_FireCostStudy_TechnicalReport.pdf.

5. Ross Gorte, "The Rising Cost of Wildfire Protection" (Headwaters Economics,
June 2013), https://headwaterseconomics.org/wp-content/uploads/fire-costs
-background-report.pdf.

6. "Colorado Wildfires: State and Private Lands," Colorado State Forest Service, http://www.leg.state.co.us/CLICS/CLICS2013A/commsumm.nsf/b4a39

62433b52fa787256e5f00670a71/789014afd43d5ebd87257bc80051a1ea/$FILE
/130815%20AttachE.pdf.

7. Headwaters Economics, "As Wildland Urban Interface (WUI) Develops, Fire-fighting Costs Will Soar," February 2013, https://headwaterseconomics.org
/dataviz/wui-development-and-wildfire-costs/.

13. PLAYING WITH FIRE

1. Elizabeth Hernandez, "New Count Shows Lefthand Canyon with the Heaviest Cyclist Use in Boulder County," *Boulder Daily Camera*, April 16, 2014, http:
//www.dailycamera.com/boulder-county-news/ci_25581681/new-count
-shows-lefthand-canyon-heaviest-cyclist-use.

2. Tom Ragan, "Fire Department: Golfer Starts 12-Acre Blaze," *Los Angeles Times*, August 30, 2010, http://www.latimes.com/tn-dpt-0831-fire-20100830-story.
html.

3. Randy Youngman, "How Hot Was It at Arroyo Trabuco?," Orange County (CA) Register, July 22, 2011, http://www.ocregister.com/articles/parsons
-305539-hot-golf.html.

4. Bill Gabbert, "Golf Club May Have Started Poinsettia Fire in California," Wildfire Today, October 19, 2014, http://wildfiretoday.com/2014/10/19/golf-club
-may-have-started-poinsettia-fire-in-california/.

5. Pat Reavy and Jeff Finley, "Nearly 9,000 Residents Evacuate Neighborhoods Near Saratoga Springs Fire," Deseret News, June 22, 2012, http://www.deseret
news.com/article/865557959/Evacuations-ordered-near-Saratoga-Springs
-fire.html.

6. Cimaron Neugebauer, "Two Men Charged for Allegedly Starting the Dump Fire in Utah County," *Salt Lake Tribune*, October 18, 2012, http://archive.sltrib
.com/story.php?ref=/sltrib/news/55109351-78/fire-shooting-target-sltrib.html.
csp.

7. Bill Gabbert, "Exploding Targets, an Increasing Wildfire Problem," Wildfire Today, October 11, 2012, http://wildfiretoday.com/2012/10/11/exploding-tar
gets-an-increasing-wildfire-problem/.

8. Bill Gabbert, "Update on the Legality of Sky Lanterns—Banned in 29 States," Wildfire Today, December 31, 2015, http://wildfiretoday.com/2015/12/31/up
date-on-the-legality-of-sky-lanterns-banned-in-28-states/.

14. NUCLEAR FRYING PAN

1. Robert Alvarez and Joni Arends, "Fire, Earth and Water: An Assessment of the Environmental, Safety and Health Impacts of the Cerro Grande Fire on the Los Alamos National Laboratory, a Department of Energy Facility" (Concerned Citizens for Nuclear Safety and the Nuclear Policy Project, December 2000), http://www.nuclearactive.org/docs/CerroGrandeindex.html.

2. Kathleene Parker, "Predictable Disaster: Ten Years After the Cerro Grande

Fire," *Forest Magazine* (Forest Service Employees for Environmental Ethics), Spring 2010.

3. Ibid.
4. J. Marshak, E. Teller, and L. R. Klein, "Dispersal of Cities and Industries," *Bulletin of the Atomic Scientists,* 1, no. 9 (April 15, 1946).
5. Roger Kennedy, *Wildfire and Americans: How to Save Lives, Property, and Your Tax Dollars* (New York: Hill and Wang, 2007).
6. Ibid.
7. Ibid.
8. Ibid.
9. Ibid.
10. Sergiy Zibtsev et al., "Wildfires Risk Reduction from Forests Contaminated by Radionuclides: A Case Study of the Chernobyl Nuclear Power Plant Exclusion Zone" (paper presented at the Fifth International Wildland Fire Conference, Sun City, South Africa, May 9–13, 2011).
11. Ron Broglio, "The Creatures That Remember Chernobyl: Radioactive Boars and Bunnies Won't Let Us Forget About the Nuclear Disaster," *Atlantic,* August 26, 2016.
12. Timothy A. Mousseau et al., "Highly Reduced Mass Loss Rates and Increased Litter Layer in Radioactively Contaminated Areas," *Oecologia* 175, no. 1 (2014): 429, doi:10.1007/s00442-014-2908-8.
13. Nikolaos Evangeliou et al., "Wildfires in Chernobyl-Contaminated Forests and Risks to the Population and Environment, *Environment International* 73 (2014): 346–58.
14. Stephen Lendman, "Forest Fires in Ukraine: Chernobyl All Over Again," Global Research, April 29, 2015, http://www.globalresearch.ca/forest-fires-in-ukraine-chernobyl-all-over-again/5446108.
15. LeRoy Moore, "LeRoy Moore: Rocky Flats Burn a Bad Idea," *Boulder Daily Camera,* November 21, 2014, http://www.dailycamera.com/guest-opinions/ci_26988064/leroy-moore-rocky-flats-burn-bad-idea.

15. THE VANISHING FOREST

1. "Las Conchas: Incident Overview," InciWeb, Incident Information System, Santa Fe National Forest, June 26, 2011, last updated June 2, 2013, https://inciweb.nwcg.gov/incident/2385.
2. "The Las Conchas Fire," Bandelier National Monument, New Mexico, https://www.nps.gov/band/learn/nature/lasconchas.htm.
3. Mary Caperton Morton, "Fire-Driven Clouds and Swirling Winds Whipped Up Record-Setting New Mexico Blaze," *Earth,* April 12, 2015, http://www.earthmagazine.org/article/fire-driven-clouds-and-swirling-winds-whipped-record-setting-new-mexico-blaze.
4. Christina L. Tague, Nathan G. McDowell, and Craig D. Allen, "An Integrated Model of Environmental Effects on Growth, Carbohydrate Balance, and Mor-

tality of *Pinus ponderosa* Forests in the Southern Rocky Mountains," *PLOS ONE* 8, no. 11 (2013), doi:10.1371/journal.pone.0080286.

5. David M.J.S. Bowman et al., "Abrupt Fire Regime Change May Cause Landscape Wide Loss of Mature Obligate Seeder Forests," *Global Change Biology* 20, no. 3 (2014): 1008–15, doi:10.1111/gcb.12433, http://onlinelibrary.wiley.com /doi/10.1111/gcb.12433/full.

6. Thomas W. Swetnam et al., "Multiscale Perspectives of Fire, Climate and Humans in Western North America and the Jemez Mountains, USA," *Philosophical Transactions of the Royal Society B* 371, no. 1696 (June 5, 2016), doi:10.1098 /rstb.2015.0168.

7. Christopher I. Roos and Thomas W. Swetnam, "A 1416-Year Reconstruction of Annual, Multidecadal, and Centennial Variability in Area Burned for Ponderosa Pine Forests of the Southern Colorado Plateau Region, Southwest USA," *Holcene* 22, no. 3 (2011): 281–90.

8. John Fleck, "New Effort Aims to Protect Watersheds," *Albuquerque Journal*, January 4, 2014, https://www.abqjournal.com/521019/new-effort-aims-to -protect-watersheds.html.

9. A. Park Williams et al., "Temperature as a Potent Driver of Regional Forest Drought Stress and Tree Mortality," *Nature Climate Change* 3 (2013): 292–97, doi:10.1038/nclimate1693.

10. U.S. Department of Agriculture, "New Aerial Survey Identifies More Than 100 Million Dead Trees in California," news release, November 18, 2016, https: //www.fs.fed.us/news/releases/new-aerial-survey-identifies-more-100-mil lion-dead-trees-california.

11. Kurtis Alexander, "An Eye-Opening Flight over California's Dying Forests," *San Francisco Chronicle*, August 6, 2016, http://www.sfchronicle.com/bayarea /article/A-hair-raising-flight-over-California-s-dying-9127084.php.

12. Jeffrey A. Hicke et al., "Effects of Biotic Disturbances on Forest Carbon Cycling in the United States and Canada," *Global Change Biology* 18 (2012): 7–34.

13. Jose Iniguez, "Whitewater Baldy Fire" (Southwest Fire Science Consortium, November 2012), http://swfireconsortium.org/wp-content/uploads/2012/11 /Whitewater_Baldy_Iniguez_bsw_reduced.pdf.

14. Molly E. Hunter, Jose M. Iniguez, and Calvin A. Farris, "Historical and Current Fire Management Practices in Two Wilderness Areas in the Southwestern United States" (U.S. Forest Service, Rocky Mountain Research Station, General Technical Report RMRS-GTR-325, August 2014), https://www.fs.fed.us /rm/pubs/rmrs_gtr325.pdf.

16. THE FIRE-INDUSTRIAL COMPLEX

1. James E. Hubbard, "2012 Wildfire Guidance" (U.S. Forest Service, May 25, 2012), https://www.documentcloud.org/documents/407523-2012-wildfire -guidance-memo-may-25.html.

2. U.S. Forest Service, "The Rising Cost of Wildfire Operations: Effects on the

Forest Service's Non-Fire Work" (August 4, 2015), https://www.fs.fed.us/sites /default/files/2015-Fire-Budget-Report.pdf.

3. Phil Taylor, "'It's Just Nuts' as Wildfires Drain Budgets Once Again," *Greenwire*, E&E News, October 30, 2013, http://www.eenews.net/stories/1059989688.

4. Bill Gabbert, "New Mexico Fire Becomes Largest in State History," Wildfire Today, May 30, 2012, http://wildfiretoday.com/2012/05/30/new-mexico-fire -becomes-largest-in-state-history/.

5. "Quadrennial Fire Review 2009" (final report, U.S. Departments of the Interior and Agriculture and National Association of State Foresters, January 2009), https://www.forestsandrangelands.gov/strategy/documents/foundational /qfr2009final.pdf.

6. Robert W. Mutch, "Wilderness Experience Right in Our Backyard Offers Lesson in Fires," Missoulian, October 10, 2012, http://missoulian.com/news/opinion /columnists/wilderness-experience-right-in-our-backyard-offers-lesson-in -fires/article_a9a9e7cc-12e3-11e2-86a9-001a4bcf887a.html.

7. Steven Pearce, "Managing Our National Forests," *Congressional Chronicle*, C-SPAN Video Library, June 21, 2012, http://www.c-spanvideo.org/video Library/clip.php?appid=601883050.

8. "Company History," National Wildfire Suppression Association, http://nwsa .publishpath.com/nwsa-history.

9. William A. Derr for Steve Pearce, "Wildfire Review Report: Whitewater Baldy Complex and Little Bear Fires; New Mexico, May thru June 2012," https: //pearce.house.gov/sites/pearce.house.gov/files/Derr%20Fire%20Report .pdf.

17. HIGH PARK

1. "Family Reported Fire, Watched It Smolder for Hours," 9News, KUSA, Denver, June 12, 2012.

2. "Firefighter Continues to Battle High Park Fire After Losing His Home," CBS4, Denver, http://denver.cbslocal.com/2012/06/13/firefighter-continues-to-battle -blaze-after-losing-his-home/.

3. Max A. Moritz et al., "Climate Change and Disruptions to Global Fire Activity," *Ecosphere* 3, no. 6 (2012): 49, http://dx.doi.org/10.1890/ES11-00345.1.

4. Katharine Hayhoe, "Climate Scientist, Christian on Climate Change and Wild-fires," Sojourners, June 22, 2012, https://sojo.net/articles/climate-scientist -christian-climate-change-and-wildfires.

5. Kate Sheppard, "Newt Dumps Christian Climate Scientist," Mother Jones, January 6, 2012, http://www.motherjones.com/environment/2012/01/newt -dumps-leading-climate-scientist.

6. Anthony Leiserowitz et al., "Climate Change in the Texas Mind" (Yale Project on Climate Change Communication, 2013), http://environment.yale.edu /climate-communication-OFF/files/Texas_Climate_Change_Report.pdf.

7. Donald Wuebbles et al., "CMIP5 Climate Model Analyses: Climate Extremes in the United States," *Bulletin of the American Meteorological Society* 95, no. 4 (April 2014): 571–83.

8. David M. Romps et al., "Projected Increase in Lightning Strikes in the United States Due to Global Warming," *Science* 346, no. 6211 (November 14, 2014): 851–54, doi: 10.1126/science.1259100.

9. X. Yue et al., "Ensemble Projections of Wildfire Activity and Carbonaceous Aerosol Concentrations over the Western United States in the Mid-21st Century," *Atmospheric Environment* 77 (October 2013), https://www.ncbi.nlm.nih.gov/pubmed/24015109.

10. W. Matt Jolly et al., "Climate-Induced Variations in Global Wildfire Danger from 1979 to 2013," *Nature Communications* 6 (July 2015), doi:10.1038/ncomms8537.

11. Caroline Perry, "Wildfires Projected to Worsen with Climate Change," *Harvard Gazette,* August 28, 2013, http://news.harvard.edu/gazette/story/2013/08/wildfires-projected-to-worsen-with-climate-change/.

12. Chris Mooney, "A Stunning Five Million Acres Have Now Burned in Alaskan Wildfires This Year," *Washington Post,* August 10, 2015, https://www.washingtonpost.com/news/energy-environment/wp/2015/08/10/five-million-acres-have-now-burned-in-alaskan-wildfires-this-year/?utm_term=.a11281ed798c.

13. "Alaska Entering New Era for Wildfires," Climate Central, June 24, 2015, http://www.climatecentral.org/news/alaska-entering-new-era-for-wildfires-19146.

14. Michelle C. Mack et al., "Carbon Loss from an Unprecedented Arctic Tundra Wildfire," *Nature* 475 (July 28, 2011), http://www.nature.com/nature/journal/v475/n7357/full/nature10283.html.

15. David M.J.S. Bowman et al., "Fire in the Earth System," *Science* 324, no. 5926 (April 24, 2009): 481–84, doi: 10.1126/science.1163886.

18. FIREBUGS

1. Jessica Fender, "Ten from Shambhala Community Dig In, Protect Stupa from High Park Fire," *Denver Post,* June 15, 2012, http://www.denverpost.com/2012/06/15/ten-from-shambhala-community-dig-in-protect-stupa-from-high-park-fire/.

2. "Wolves Protected During Fire Evacuations," CBS4, Denver, June 14, 2012, http://denver.cbslocal.com/2012/06/14/wolves-protected-during-fire-evacuations/#.T9vYnFCaCD0.twitter.

3. "Vets Think Donkey Minding Draft Horses Did Pasture Heroics During Fire," *Denver Post,* June 15, 2012, http://www.denverpost.com/2012/06/15/vets-think-donkey-minding-draft-horses-did-pasture-heroics-during-fire/.

4. "June 22 to June 28: This Week in Denver Weather History," ThorntonWeather.com, June 26, 2014, http://www.thorntonweather.com/blog/weather-history/june-22-to-june-28-this-week-in-denver-weather-history/.

5. "Crews Cede Ground to High Park Fire," CBS4, Denver, June 23, 2012, http://denver.cbslocal.com/2012/06/23/crews-cede-ground-to-northern-colorado-wildfire/.

6. Shane Benjamin, "Wind-Whipped Fire Temporarily Closes Highway 550 Near Bondad," *Durango (CO) Herald,* June 24, 2012, https://durangoherald.com/articles/40533-wind-whipped-fire-temporarily-closes-highway-550-near-bondad.

7. "Treasure Fire Burns Between Leadville and Alma," *Aspen (CO) Times,* June 25, 2012, http://www.aspentimes.com/news/treasure-fire-burns-between-leadville-and-alma/.

8. Associated Press, "Coal Mine Smoldering Poses Fire Risk in SW Colo.," *Denver Post,* May 7, 2013, http://www.denverpost.com/2013/05/07/coal-mine-smoldering-poses-fire-risk-in-sw-colo/.

9. El Paso County Sheriff's Office, "Waldo Canyon Fire After Action Report" (Colorado Springs, CO, April 19, 2013), http://wildfiretoday.com/documents/Waldo_Canyon_Fire_Sheriff_Report.pdf.

10. Cassandra Profita, "Court: Yellowstone Grizzlies Still Need Protection," *Ecotrope,* Oregon Public Broadcasting, November 22, 2011, updated February 19, 2013, http://www.opb.org/news/blog/ecotrope/court-yellowstone-grizzlies-still-need-protection-as-beetle-outbreaks-kill-a-key-food-source/.

11. Bill Gabbert, "Bark Beetles and Wildland Fire Behavior — A Summary of Research," Wildfire Today, June 1, 2012, http://wildfiretoday.com/2012/06/01/bark-beetles-and-wildland-fire-behavior-a-summary-of-research/.

12. Matt Jolly et al., "Mountain Pine Beetle Effects on Fire Behavior," U.S. Forest Service, 2011, https://www.fs.fed.us/research/highlights/highlights_display.php?in_high_id=395.

13. Gail Wells, "Bark Beetles and Fire: Two Forces of Nature Transforming Western Forests," *Fire Science Digest* (Joint Fire Science Program), 12 (February 2012).

14. Jeffry B. Mitton, Scott M. Ferrenberg, and Craig W. Benkman, "Mountain Pine Beetle Develops an Unprecedented Summer Generation in Response to Climate Warming," *American Naturalist* 179, no. 5 (2012): E163–71, doi:10.1086/665007.

15. Jennifer K. Balch et al., "Introduced Annual Grass Increases Regional Fire Activity Across the Arid Western USA (1980–2009)," *Global Change Biology,* 19 (2013): 173–83, doi:10.1111/gcb.12046.

16. "Invasive Cheatgrass (*Bromus tectorum*)," University of Wisconsin–La Crosse, http://bioweb.uwlax.edu/bio203/s2014/klein_shan/.

19. FOREST JIHAD

1. Randy Kreider, "Al Qaeda Magazine Calls for Firebomb Campaign in US," ABC News, May 2, 2012, http://abcnews.go.com/Blotter/al-qaeda-calls-massive-forest-fires-montana/story?id=16263981.

2. Diane French, "Forests Under Fire: Religious-Inspired Pyro-Terrorism in U.S.

Forests" (research paper, International Forestry course, Johns Hopkins University, n.d.).

3. Miriam Elder, "Russia Accuses al-Qaida of 'Forest Jihad' in Europe," *Guardian,* October 3, 2012, https://www.theguardian.com/world/2012/oct/03/russia-al-qaida-forest-jihad.

4. Jerry Williams et al., "Findings and Implications from a Course-Scale Global Assessment of Recent Selected Mega-Fires" (paper presented at the Fifth International Wildland Fire Conference, Sun City, South Africa, May 9–11, 2011).

5. John Ingold, "Decade After Hayman Fire, Questions Linger About Fire's Start," *Denver Post,* June 2, 2012, http://www.denverpost.com/2012/06/02/decade-after-hayman-fire-questions-linger-about-fires-start/.

6. "Nearly 350 Homes Destroyed in Colorado Springs, Officials Say," CNN, June 28, 2012, http://www.cnn.com/2012/06/28/us/western-wildfires/.

7. Jeffrey T. Prestemon and David T. Butry, "Wildland Arson: A Research Assessment," in *Advances in Threat Assessment and Their Application to Forest and Rangeland Management,* vol. 1 (U.S. Forest Service, Pacific Northwest Research Station, General Technical Report PNW-GTR-802, September 2010), https://www.fs.fed.us/pnw/pubs/gtr802/Vol2/pnw_gtr802_prestemon02.pdf.

8. Jennifer K. Balch et al., "Human-Started Wildfires Expand the Fire Niche Across the United States," *Proceedings of the National Academy of Sciences* 114, no. 11 (February 27, 2017): 2946–51, http://www.pnas.org/content/early/2017/02/21/1617394114.full.

20. MOUNTAIN SHADOWS

1. Ryan Maye Handy and Daniel J. Chacon, "Waldo Canyon Fire: Homes Burn as Evacuation Calls Go Out," *Colorado Springs Gazette,* July 19, 2012.

2. Ibid.

21. FIRESTORM

1. David Olinger and Jeremy P. Meyer, "Tapes Show Waldo Canyon Fire Evacuations Delayed Two Hours," *Denver Post,* July 19, 2012.

2. Daniel Chacon, "Waldo Canyon Fire Erupts During Press Conference," YouTube, https://www.youtube.com/watch?v=x1xqRk7wgTs.

3. Olinger and Meyer, "Tapes Show."

4. Ibid.

5. City of Colorado Springs, "Waldo Canyon Fire, 23 June 2012 to 10 July 2012: Final After Action Report" (April 3, 2013), https://www.springsgov.com/units/communications/ColoradoSpringsFinalWaldoAAR_3April2013.pdf.

6. Ibid.

7. Ryan Maye Handy, "Waldo Canyon Fire: Firefighters Defended Streets They Had Never Known," *Colorado Springs Gazette,* July 7, 2012.

8. Ibid.

9. "Colorado Wildfires Escalate into Major Disaster," Colorado Pols, June 27, 2012, http://www.coloradopols.com/diary/18031/colorado-wildfires-escalate -into-major-disaster#sthash.D5rBflfO.kTj4l0hB.dpbs.

22. SEEING RED

1. Bill Gabbert, "Two Air Tanker Incidents, One Crash and One Wheels-Up Landing," Wildfire Today, June 3, 2012, http://wildfiretoday.com/2012/06/03/two -air-tanker-incidents-one-crash-and-one-wheels-up-landing/.
2. Bill Gabbert, "20 Fatalities in P2V Air Tanker Crashes Since 1987," Wildfire Today, June 7, 2012, http://wildfiretoday.com/2012/06/07/24-fatalities-in-p2v -air-tanker-crashes-since-1974/.
3. "Federal Aerial Firefighting: Assessing Safety and Effectiveness; Blue Ribbon Panel Report to the Chief, USDA Forest Service, and Director, USDI Bureau of Land Management" (December 2002), https://www.fs.fed.us/fire/publica tions/aviation/fed_aerial_ff_assessing_safety_effectivenss_brp_2002.pdf.
4. Brian Skoloff and Scott Sonner, "'The Air Tanker Fleet Continues to Atrophy,'" Associated Press, CBS13, KVAL, Denver, June 3, 2012, http://kval.com/out doors/the-air-tanker-fleet-continues-to-atrophy.
5. Randall Stephens, "Is Washington Policy Conducive to a Stable Civil AFF Industry?" *Fireplanes* (blog), May 30, 2012, https://fireplanes.wordpress. com/2012/05/30/washington-will-let-americans-die-in-2012-wildfires-due -to-attention-deficits/.
6. Michelle Malkin, "How Obama Bureaucrats Fueled Western Wildfires," MichelleMalkin.com, June 20, 2012, http://michellemalkin.com/2012/06/20 /how-obama-bureaucrats-fueled-western-wildfires/.
7. "National Guard Firefighting Planes Sit Idle as Obama Authorizes $24 Million for New Aircraft," Associated Press, FoxNews.com, June 15, 2012, http://www .foxnews.com/us/2012/06/14/as-wildfires-rage-modern-tanker-planes -sought.html.
8. White House, "Readout of President Obama Signing S. 3261, Contract for Large Air Tankers," news release, June 13, 2012, https://www.whitehouse.gov/the -press-office/2012/06/13/readout-president-obama-signing-s-3261-contract -awards-large-air-tankers.
9. Ibid.
10. Jeremy P. Meyer, "Officials Disagree on Ability of Nation's Old, Thin Air Tanker Fleet," *Denver Post,* June 16, 2012, http://www.denverpost.com/2012/06/16 /officials-disagree-on-ability-of-nations-old-thin-air-tanker-fleet/.
11. Bill Gabbert, "Air Tankers in the News," Wildfire Today, June 15, 2012, http://wildfiretoday.com/2012/06/15/air-tankers-news/.
12. "Modular Airborne Fire Fighting Systems (MAFFS)," Fire and Aviation Management, U.S. Forest Service, https://www.fs.fed.us/fire/aviation/airplanes /maffs.HTML.
13. Michael Collins, "Debate over Use of MAFFS an Old Battle," *Ventura*

County Star (Camarillo, CA), July 9, 2012, http://www.military.com/daily
-news/2012/07/09/debate-over-use-of-maffs-an-old-battle.html.

14. Herbert E. McLean, "The Great Air Tanker Debacle," *American Forests,* May 1, 1994.

15. "Only the Godfather: CIA and Subsidiaries Exposed in Court Documents as Active Drug Smugglers Using Military Aircraft Washed Through Forest Service," *From the Wilderness Newsletter,* December 1998, http://www.fromthewilder ness.com/free/pandora/forest_service_c130s.html.

16. Bill Gabbert, "Summary of the Investigation into the Iron Complex Fire, 9-Fatality Helicopter Crash," Wildfire Today, December 8, 2010, http://wildfire today.com/2010/12/08/summary-of-the-investigation-into-the-iron-44-fire -9-fatality-helicopter-crash/.

17. Chuck Sheley, "The Selected Few and 'The List,'" *Smokejumper,* July 2015, http://smokejumpers.com/index.php/smokejumpermagazine/getitem/arti cles_id=396.

18. Nichole Grady and Ann Skarban, "Fire's Fury Presents Challenges to MAFFS Operationally, Personally," 302nd Airlift Wing, June 27, 2012, http: //www.302aw.afrc.af.mil/News/ArticleDisplay/tabid/2343/Article/190252 /fires-fury-presents-challenges-to-maffs-operationally-personally.aspx.

19. Ann Skarban, "MAFFS 2012: Busy Year for 302nd Includes Battling Fires in Own 'Backyard,'" *Citizen Airman,* September 27, 2012, http://www.citamn.afrc .af.mil/News/Features/tabid/8673/Article/195113/maffs-2012-busy-year-for -302nd-includes-battling-fires-in-own-backyard.aspx.

20. "United States Air Force Aircraft Accident Investigation Board Report: C-130H3, T/N 93-1458" (North Carolina Air National Guard, Charlotte, NC, November 2, 2012).

21. Ibid.

22. U.S. Government Accountability Office, "Wildland Fire Management: Improvements Needed in Information, Collaboration, and Planning to Enhance Federal Fire Aviation Program Success" (Washington, DC, August 2013), http://wild firetoday.com/documents/2013_GAO_Air_Tanker_Study.pdf.

23. David E. Calkin et al., "Large Airtanker Use and Outcomes in Suppressing Wildland Fires in the United States," *International Journal of Wildland Fire* 23 (2014): 259–71, http://wildfiretoday.com/documents/LAT.pdf.

24. "'It Was a Red River': Fire Retardant Drop Kills Fall River Fish, Reaches Deschutes," *Bend (OR) Bugle,* August 2002, https://www.bendbugle.com/2002/08/it -was-a-red-river-fire-retardant-drop-kills-fall-river-fish-reaches-deschutes/.

25. Bruce Finley, "Wildfire: Red Slurry's Toxic Dark Side," *Denver Post,* June 16, 2012, http://www.denverpost.com/2012/06/16/wildfire-red-slurrys-toxic -dark-side/.

26. Julie Cart and Bettina Boxall, "Air Tanker Drops in Wildfires Are Often Just for Show," *Los Angeles Times,* July 29, 2008, http://www.latimes.com/local/la-me -wildfires29-2008jul29-story.html.

27. G. Sam Foster et al., "Serious Accident Investigation Report: Steep Corner Fire

Fatality" (North Fork Ranger District, Clearwater National Forest, Idaho, January 2, 2013), http://wildfiretoday.com/documents/Steep_Corner_Fatality_SAI.pdf.

28. Ibid.

29. SAFENET, Wildland Fire Safety and Health Reporting Network, "Steep Corner" (ID # 8X6NAGSAFE, August 14, 2012), http://safenet.nifc.gov/view_safenet.cfm?id=25666.

30. Ibid.

31. Foster et al., "Serious Accident Investigation Report."

32. U.S. Forest Service, "Forest Service 2016 Wildland Fire Risk Management Protocols," https://www.fs.fed.us/r1/fire/nrcg/wfdss/NR%20Support%2Docs/2016%20Fire%20Risk%20Management%20Protocols.pdf.

33. Betsy Z. Russell, "Wildland Firefighter's Death Raises Safety Questions," *Spokesman Review* (Spokane, WA), August 22, 2012, http://www.firehouse.com/news/10762682/wildland-firefighters-death-raises-safety-questions.

34. Foster et al., "Serious Accident Investigation Report."

35. Bill Gabbert, "OSHA Revises Fine and Citation for the Steep Corner Fire Fatality," Wildfire Today, March 22, 2013, http://wildfiretoday.com/2013/03/22/osha-revises-fine-and-citation-for-the-steep-corner-fire-fatality/.

36. Nicholas K. Geranios, "'I Felt Like My Heart Ripped Apart': Mother of Fallen Female Firefighter, 20, Killed in Wildfire Speaks of Devastation," Associated Press, *Daily Mail*, August 17, 2012, http://www.dailymail.co.uk/news/article-2190126/US-Wildfires-Mother-fallen-female-firefighter-killed-wildfires-speaks-devastation.html.

23. NEVER WINTER

1. "Wildfires — August 2012," State of the Climate, National Centers for Environmental Information, https://www.ncdc.noaa.gov/sotc/fire/201208.

2. Betsy Z. Russell, "Fires Intensify, Rafters Shuttled Out from Middle Fork of Salmon River," *Eye on Boise* (blog), *Spokesman Review* (Spokane, WA), August 15, 2012, http://www.spokesman.com/blogs/boise/2012/aug/15/fires-intensify-rafters-shuttled-out-middle-fork/.

3. Kelly Andersson, "Oklahoma Firefighters Scrambling," Wildfire Today, August 3, 2012, http://wildfiretoday.com/2012/08/03/oklahoma-firefighters-scrambling/.

4. "Emporia State Student-Athlete Pays It Forward," Emporia State University News, March 1, 2013, https://www.emporia.edu/news/03/01/2013/emporia-state-student-athlete-pays-it-forward/?.

5. "Wildfire Rips Through Miles of Central Wash.," CBS News/Associated Press, August 14, 2012, http://www.cbsnews.com/news/wildfire-rips-through-miles-of-central-wash/.

6. "Wildfire Shuts Down Pony Express Road," *Ogden (UT) Standard-Examiner*, August 13, 2012, http://www.standard.net/Local/2012/08/13/Wildfire-shuts-down-Pony-Express-Road.

7. Ronnie Cohen, "Western Wildfires Force Evacuations, Firefighter Killed," Reuters, *Chicago Tribune,* August 13, 2012, http://articles.chicagotribune .com/2012-08-13/news/sns-rt-us-usa-california-wildfirebre87 c0zr-20120813_1_forestry-and-fire-protection-fire-management-team-daniel-berlant.

8. Bill Gabbert, "Old Fire Arsonist Receives Death Penalty," Wildfire Today, January 28, 2013, http://wildfiretoday.com/2013/01/28/old-fire-arsonist-receives -death-penalty/.

9. "Fern Lake Fire Offers Challenges, Opportunities: Rocky Mountain National Park, Colorado Cohesive Strategy — Response to Wildfire," *Fire Stories,* Fire and Aviation Management, National Park Service, https://www.nps.gov/fire/wild land-fire/connect/fire-stories/2012-parks/rocky-mountain-national-park-3 .cfm.

24. BLACK FOREST

1. Bruce Finley, "West Fork Fire Complex in Colorado Feeding on Beetle-Ravaged Forests," *Denver Post,* June 22, 2013, http://www.denverpost.com/2013/06/22 /west-fork-fire-complex-in-colorado-feeding-on-beetle-ravaged-forests/.

2. Jeri Clausing, "Firefighters Keep Guard over Mountain Enclave," Associated Press, June 22, 2013, https://www.usnews.com/news/us/articles/2013/06/22 /firefighters-keep-guard-over-mountain-enclave.

3. Ryan Maye Handy, "Investigation into Waldo Canyon Fire Has Stalled, Detective Says," *Colorado Springs Gazette,* June 7, 2013, http://gazette.com/investiga tion-into-waldo-canyon-fire-has-stalled-detective-says/article/1502054.

4. "Former Detective Reveals Black Forest Fire Theory," CBS11, KKTV, Colorado Springs, October 19, 2016, http://www.kktv.com/content/news/Former-detec tive-reveals-Black-Forest-Fire-theory-397688431.html.

5. Alison Noon, "Complete Timeline of Black Forest Fire Events," *Colorado Springs Gazette,* June 15, 2013, http://gazette.com/complete-timeline-of-black-forest -fire-events/article/1502358.

6. David Fisher, "Black Forest Fire/Rescue Protection District Investigative Report" (Fisher Enterprises LLC, Black Forest Fire/Rescue Protection District, March 14, 2014).

7. Ibid.

8. Micholas Ricardi, "Neighbors: Couple Killed in Black Forest Fire Waited for Order to Leave," Associated Press, *Colorado Springs Gazette,* June 19, 2013, http: //gazette.com/neighbors-couple-killed-in-black-forest-fire-waited-for-order -to-leave/article/1502489.

9. Fisher, "Black Forest Fire."

10. Ibid.

11. SAFENET, Wildland Fire Safety and Health Reporting Network, "Black Forest Fire" (ID # 20131015-0001, October 15, 2013), http://safenet.nifc.gov/view _safenet.cfm?id=30121&__ncforminfo=Umo1cfXEb0lDGme9HeeAyZNxlAds

YIbYHGZ8WvqTvNCjFFLW3HIRahuGJJpQgtwUZ0OouCKQqEKC5W3khr
FaI7wHiGAYiTXn.

12. Fisher, "Black Forest Fire."

13. Ryan Parker, "Black Forest Fire: Report Says Crew Diverted, Sheriff Calls It 'Garbage,'" *Denver Post,* March 14, 2014, http://www.denverpost.com/2014/03/14
/black-forest-fire-report-says-crew-diverted-sheriff-calls-it-garbage/.

14. El Paso County Sheriff's Office, "Black Forest Fire, 11–21 June 2013, After Action
Report/Improvement Plan" (Colorado Springs, May 15, 2014), http://wildfire
today.com/documents/Black_Forest_Fire_EPSO_AA_Report.pdf.

25. TRICKLE DOWN

1. Susan Cannon and Joseph Gartner, "Wildfire-Related Debris Flow from a Hazards Perspective," in Matthias Jakob and Oldrich Hungr, *Debris-Flow Hazards
and Related Phenomena* (Chichester, UK: Springer, 2005).

2. Jenny Deam, "A Year After Waldo Canyon Fire, Colorado Town Contends with
Flooding," *Los Angeles Times,* August 29, 2013, http://articles.latimes.com/2013
/aug/29/nation/la-na-manitou-springs-20130830.

3. "Preliminary Evaluation of the Fire-Related Debris Flows on Storm King Mountain, Glenwood Springs, Colorado" (U.S. Geological Survey, 1995), https://pubs
.usgs.gov/of/1995/ofr-95-0508/skrep2.html.

4. Miguel Bustillo, "Fires Cleared Way for Debris Flows: The San Bernardino
Area Is Especially Vulnerable Because of Residential Growth," *Los Angeles
Times,* December 28, 2003.

5. Joseph Serna and Emily Foxhall, "Thermometer Hits Triple Digits as Heat
Wave Strikes Southern California," *Los Angeles Times,* June 28, 2013, http:
//articles.latimes.com/2013/jun/28/local/la-me-ln-la-heat-wave-20130628.

6. Becky Oskin, "How California's Rim Fire Grew So Big," *Planet Earth,* Live Science, September 4, 2013, http://www.livescience.com/39408-how-rim-fire
-grew-big.html.

7. Katie Valentine, "California's Rim Fire Threatens San Francisco's Power and
Water Supply," ThinkProgress, August 26, 2013, https://thinkprogress.org
/californias-rim-fire-threatens-san-francisco-s-power-and-water-supply-da
0331cac131#.gk8mesy7o.

8. Chris Roberts, "Rim Fire Damage Costs San Francisco $36.3 million," *San Francisco Examiner,* November 13, 2013, http://archives.sfexaminer.com/sanfrancisco/rim-fire-damage-costs-san-francisco-363-million/Content?oid=2625731.

9. Alan Fram, "Budget Cuts Trim Federal Wildfire Spending," Associated Press,
Las Vegas Sun, July 3, 2013.

10. Bill Gabbert, "Secretaries Vilsac and Jewell Discuss Wildfire Preparedness,"
Wildfire Today, May 13, 2013, http://wildfiretoday.com/2013/05/13/secretaries
-vilsack-and-jewell-discuss-wildfire-preparedness/.

11. U.S. Forest Service, "The Rising Cost of Wildfire Operations: Effects on the

Forest Service's Non-Fire Work" (August 4, 2015), https://www.fs.fed.us/sites /default/files/2015-Fire-Budget-Report.pdf.

12. U.S. Department of Agriculture, "Cost of Fighting Wildfires in 2014 Projected to Be Hundreds of Millions of Dollars over Amount Available," news release, May 1, 2014, https://originwww.fs.fed.us/news/releases/cost-fighting-wild fires-2014-projected-be-hundreds-millions-dollars-over-amount.

13. U.S. Department of Agriculture, "New Report Shows Budget Impact of Rising Firefighting Costs, Other Forest Programs, Including Efforts to Help Prevent and Mitigate Fire Damage Have Shrunk; Secretary Vilsack Renews Call to Better Protect Public Forests from Wildfire Threats," news release, August 20, 2014, https://www.usda.gov/wps/portal/usda/usdahome?contentid=2014/08/0184 .xml.

14. Raju Chebium, "Report Shows Impact of 'Fire Borrowing' on Forest Service," Gannett Washington Bureau, August 20, 2014, http://www.azcentral.com /story/news/politics/2014/08/20/vilsack-report-fire-borrowing-impact-for est-service/14359173/.

15. U.S. Senate, Committee on Energy and Natural Resources, letter to Sylvia Burwell, Tom Vilsack, and Sally Jewell, June 28, 2013, https://www.energy .senate.gov/public/index.cfm/files/serve?File_id=a797bf00-f421-414d-8fca -bdcfe1481728.

16. National Association of State Foresters, "Timeline of Suppression, FLAME, Transfers" (February 2014), http://www.stateforesters.org/sites/default/files /publication-documents/Timeline_Feb2014.pdf.

17. Bruce Finley, "Feds Predict Climate Change Will Double Wildfire Risk in Forests," *Denver Post*, April 3, 2013, http://www.denverpost.com/2013/04/03/feds -project-climate-change-will-double-wildfire-risk-in-forests/.

26. FRONTIER DAYS

1. Andrea Aker, "The Story of Buckey O'Neill: Arizona's Happy Warrior," Arizona Oddities, July 13, 2012, http://arizonaoddities.com/2012/07/the-story-of -buckey-oneill-arizonas-happy-warrior/.

2. Richard Gorby, "Prescott's Great Fire, July 14, 1900," Sharlot Hall Museum, July 1998, https://sharlot.org/1998/july/863-prescott-s-great-fire-july-14-1900.

3. "Conflagration: The Sherman Block Reduced to Ashes — Involving the Loss of Human Life and over $50,000 Worth of Property, Fearful Death of S. N. Holmes . . . ," *Weekly Arizona Miner*, February 15, 1884.

4. "A Disastrous Fire: The East Side of the Plaza Laid in Ashes, About $50,000 Worth of Property Destroyed," *Arizona Weekly Journal-Miner*, July 2, 1888.

5. "Prescott Is Visited by a Great Fire," *Arizona Weekly Journal-Miner*, July 15, 1900.

6. Richard Gorby, "Days Past: Palace Saloon Emerged from Great Fire of 1900 Grander Than Ever," *Prescott Daily Courier*, September 29, 2012, http://www

.dcourier.com/news/2012/sep/29/days-past-palace-saloon-emerged-from
-great-fire-o/.

27. DEFUSING THE TIME BOMB

1. "Granite Mountain Hot Shots," n.d., history provided by Prescott Fire Department, Prescott, AZ.
2. Anne Ryman and Rebekah L. Sanders, "Yarnell Hill Fire Fallen Remembered: Eric Marsh," *Arizona Republic,* July 6, 2013, http://archive.azcentral.com /news/arizona/articles/20130702yarnell-fire-eric-marsh-obit.html.
3. Amanda Marsh, "About Eric," Eric Marsh Foundation for Wildland Firefighters, http://www.ericmarshfoundation.org/about-eric/.
4. Yvonne Wingett Sanchez, "Holding On to What Was Theirs: A Granite Mountain Hotshot Widow's Story," *Arizona Republic,* June 26, 2016, http://www.az central.com/story/news/local/arizona-best-reads/2016/06/26/granite-moun tain-hotshot-widows-story/85890188/.
5. "Granite Mountain Hot Shots."
6. Ryman and Sanders, "Yarnell Hill Fire Fallen Remembered: Eric Marsh."
7. Joanna Dodder Nellans, "What Stopped the Raging Indian Fire from Burning Thousands of Prescott Homes?," *Prescott Daily Courier,* May 17, 2012, https: //www.dcourier.com/news/2012/may/17/what-stopped-the-raging-indian -fire-from-burning-/.
8. Ibid.
9. Ibid.
10. Robert Anglen, Saba Hamedy, and Kristina Goetz, "Yarnell Hill Fire Fallen Remembered: Travis Turbyfill," *Arizona Republic,* July 10, 2013, http://archive.az central.com/news/articles/20130702yarnell-fire-travis-turbyfill-obit.html.
11. Zach St. George, "Yarnell Hill Fire Fallen Remembered: Andrew Ashcraft," *Arizona Republic,* July 10, 2013, http://archive.azcentral.com/news/arizona /articles/20130705yarnell-fire-andrew-ashcraft-obit.html.
12. Ray Stern, "Brendan McDonough, Sole Survivor of Firefighting Crew, Was 'Remorseful' After 2010 Conviction for Stolen Property," *Phoenix New Times,* July 3, 2013, http://www.phoenixnewtimes.com/news/brendan-mcdonough-sole -survivor-of-firefighting-crew-was-remorseful-after-2010-conviction-for-sto len-property-6626985.
13. "Fire Shelter Deployment: Mackenzie Incident, Kingman Resource Area BLM, June 1, 1994," Wildfire Lessons Learned, http://www.wildfirelessons.net/High erLogic/System/DownloadDocumentFile.ashx?DocumentFileKey=092e0f6f -4e09-4b57-bbef-d43722e56288.
14. Andrew Freedman, "The Climate Context Behind the Deadly Arizona Wildfire," Climate Central, July 1, 2013, http://www.climatecentral.org/news/the -climate-context-behind-the-deadly-arizona-wildfire-16175.
15. "History of the Prescott National Forest," Prescott National Forest, https:

//www.fs.usda.gov/detailfull/prescott/about-forest/?cid=stelprdb5121879 &width=full.

16. Robert Anglen and Saba Hamedy, "Yarnell Hill Fire Fallen Remembered: Grant Quinn McKee," *Arizona Republic,* July 6, 2013, http://archive.azcentral.com /news/arizona/articles/20130703yarnell-fire-grant-quinn-mckee-obit.html.

17. Catherine Reagor and Zach St. George, "Yarnell Hill Fire Fallen Remembered: John Percin Jr.," *Arizona Republic,* July 6, 2013, http://archive.azcentral.com /news/arizona/articles/20130702john-percin-jr.html.

18. Anglen and Hamedy, "Yarnell Hill Fire Fallen Remembered: Grant Quinn McKee."

19. "The John J. Percin Jr. Memorial Chapter 5 Scholarship Fund," https://www .facebook.com/The-John-J-Percin-Jr-Memorial-Chapter-5-Recovery-Schol arship-Fund-551588601563694/.

28. THE DOCE

1. "Doce Fire: Incident Overview," InciWeb, Incident Information System, Prescott National Forest, June 18, 2013, last updated July 12, 2013, https://inci web.nwcg.gov/incident/3437/.

2. "1, 2, 3: A Whole Community Recovery; Yavapai County, June 18th through July 7th, 2013: An Exposition of the Doce, Yarnell Hill Wildfire, and the Fallen 19" (September 22, 2013), http://coyotecampaign.org/documents/Denny%20 Faulk%20Wildfires%202013.pdf.

3. Joanna Dodder, "Hotshots Tree Earns Magnificent 7 Honor," *Prescott Daily Courier,* April 17, 2015, http://www.dcourier.com/news/2015/apr/17/hotshots -tree-earns-magnificent-7-honor/.

4. Robert Anglen, Kristina Goetz, and Saba Hamedy, "Yarnell Hill Fire Fallen Remembered: Travis Carter," *Arizona Republic,* July 10, 2013, http://archive.az central.com/news/arizona/articles/20130702yarnell-fire-travis-carter-obit .html.

5. Anne Ryman and Scott Craven, "Yarnell Hill Fire Fallen Remembered: Anthony Rose," *Arizona Republic,* July 10, 2013, http://archive.azcentral.com /news/arizona/articles/20130702anthony-rose.html.

6. Catherine Reagor, "Yarnell Hill Fire Fallen Remembered: Joe Thurston," *Arizona Republic,* July 10, 2013, http://archive.azcentral.com/news/arizona /articles/20130703joe-thurston.html.

7. Robert Anglen, Saba Hamedy, and Kristina Goetz, "Yarnell Hill Fire Fallen Remembered: Travis Turbyfill," *Arizona Republic,* July 10, 2013, http://archive .azcentral.com/news/articles/20130702yarnell-fire-travis-turbyfill-obit.html.

8. Diana M. Náñez and Karina Bland, "Yarnell Hill Fire Fallen Remembered: Billy Warneke," *Arizona Republic,* July 10, 2013, http://archive.azcentral.com/news /arizona/articles/20130702billy-warneke.html.

9. Karina Bland, "Yarnell Hill Fire Fallen Remembered: Sean Misner," *Ari-*

zona Republic, July 10, 2013, http://archive.azcentral.com/news/arizona /articles/20130702yarnell-fire-sean-misner-obit.html.

10. Brandon Brown, "Yarnell Hill Fire Fallen Remembered: Dustin Deford," *Arizona Republic,* July 10, 2013, http://archive.azcentral.com/news/arizona /articles/20130702dustin-deford.html.

11. Saba Hamedy and Robert Anglen, "Yarnell Hill Fire Fallen Remembered: Kevin Woyjeck," *Arizona Republic,* July 10, 2013, http://archive.azcentral.com/news /arizona/articles/20130702yarnell-fire-kevin-woyjeck.html.

12. Ryan Randazzo and Lindsey Collom, "Yarnell Hill Fire Fallen Remembered: Garret Zuppiger," *Arizona Republic,* July 10, 2013, http://archive.azcentral .com/news/arizona/articles/20130703yarnell-fire-garret-zuppiger-obit .html.

13. Connie Cone Sexton and Rebecca McKinsey, "Yarnell Hill Fire Fallen Remembered: Scott Norris," *Arizona Republic,* July 6, 2013, http://archive.azcentral .com/news/arizona/articles/20130705yarnell-fire-scott-norris-obit.html.

14. Ryan Randazzo and Lindsey Collom, "Yarnell Hill Fire Fallen Remembered: Chris MacKenzie," *Arizona Republic,* July 10, 2013, http://archive.azcentral .com/news/arizona/articles/20130703christopher-mackenzie.html.

15. "Granite Mountain Hotshots," exhibit, Hotel St. Michael, Prescott, AZ, June 30, 2014.

16. Ibid.

17. Scott Craven, "Yarnell Hill Fire Fallen Remembered: Clayton Whitted," *Arizona Republic,* July 10, 2013, http://archive.azcentral.com/news/arizona /articles/20130702yarnell-fire-clayton-whitted-obit.html.

18. "Granite Mountain Hotshots," exhibit.

19. Cindy Barks, "Hotshots Earned, Cost City Millions: Payouts from Fighting Fires in Other States Helped Offset Operating Costs," *Prescott Daily Courier,* August 31, 2013, http://www.dcourier.com/news/2013/aug/31/hotshots-earned-cost -city-millions-payouts-from-f/.

20. Ibid.

21. "Standards for Interagency Hotshot Crew Operations: Annual IHC Mobilization Checklist [Granite Mountain IHC]," Southwest Coordination Center (Albuquerque, NM, April 23, 2013).

22. Chris MacKenzie, personnel records, Prescott Fire Department, Prescott, AZ.

23. "Hearing for Juliann Ashcraft Before the PSPRS Local Board," *Prescott eNews,* May 21, 2014, http://www.prescottenews.com/news/current-news/item /23552-hearing-for-juliann-ashcraft-before-the-psprs-local-board.

24. Chris MacKenzie, Granite Mountain Interagency Hot Shot Crew, National Wildfire Coordinating Group, task book for positions of Firefighter Type 1 (FFT1), Incident Commander Type 5 (ICT5), initiated April 26, 2011, final evaluator's verification by Clayton Whitted, June 25, 2013.

25. "Employee Performance Appraisal: Eric Marsh," City of Prescott, AZ, May 13, 2013.

26. Ibid.

27. "Employee Performance Appraisal: Eric Marsh," City of Prescott, AZ, July 28, 2011.

29. WHERE THE DESERT BREEZE MEETS
THE MOUNTAIN AIR

1. Jim Karels et al., "Yarnell Hill Fire, June 30, 2013: Serious Accident Investigation Report" (State of Arizona, September 23, 2013), http://www.iawfonline.org/Yarnell_Hill_Fire_report.pdf.
2. Ibid.
3. Barbara Kelso, videotaped interview by reporter John Dougherty, July 6, 2013, https://www.youtube.com/watch?v=YYWM3D6RIFQ.
4. Karels et al., "Yarnell Hill Fire."
5. National Weather Service, Flagstaff, AZ, spot weather forecast for Yarnell Hill Fire, requested by initial attack incident commander at 9:56 p.m., received at 10:07 p.m., June 28, 2012.
6. Megan Finnerty, "In Yarnell, Everyone Treasures Community," *Arizona Republic*, July 5, 2013, http://archive.azcentral.com/news/arizona/articles/20130703yarnell-everyone-treasures-community.html.
7. Peter H. Morrison and George Wooten, "Analysis and Comments on the Yarnell Hill Fire in Arizona and the Current Fire Situation in the United States" (Pacific Biodiversity Institute, July 2013), http://www.pacificbio.org/initiatives/fire/THE%20YARNELL%20HILL%20FIRE%20AND%20THE%20NEED%20FOR%20FIRE%20ADAPTED%20COMMUNITIES2013july17.pdf.
8. Mary Jo Pitzl, Brandon Loomis, and Matthew Dempsey, "In Harm's Way," *Arizona Republic*, December 8, 2013, http://archive.azcentral.com/news/wildfires/yarnell/arizona-wildfires-homes-forests-risk/.

30. THE PERFECT FIRESTORM

1. Sean Holstege and Anne Ryman, "Residents' Questions on Fire Left Unanswered," *Arizona Republic*, October 1, 2013, http://archive.azcentral.com/news/arizona/articles/20131001residents-questions-fire-left-unanswered.html.
2. Jim Karels et al., "Yarnell Hill Fire, June 30, 2013: Serious Accident Investigation Report" (State of Arizona, September 23, 2013), http://www.iawfonline.org/Yarnell_Hill_Fire_report.pdf.
3. Wildland Fire Associates, prepared for Arizona Division of Occupational Safety and Health, "Granite Mountain IHC Entrapment and Burnover Investigation: Yarnell Hill Fire — June 30, 2013" (November 2013), http://www.iawfonline.org/WildlandFireAssociatesReportYARNELL.pdf.
4. Ibid.
5. Jim Karels et al., "Yarnell Hill Fire."
6. Wildland Fire Associates, "Granite Mountain IHC Entrapment and Burnover Investigation."

7. Dispatch logs, entry 2115, June 29, 2013, Arizona Interagency Dispatch Center, Phoenix, AZ.
8. Karels et al., "Yarnell Hill Fire."
9. Wildland Fire Associates, "Granite Mountain IHC Entrapment and Burnover Investigation."
10. Ibid.
11. Dispatch logs, AZ-A1S 2013-688, "Yarnell Hill" Wildfire, June 29, 2013, 18:56:28, Arizona Interagency Dispatch Center, Phoenix, AZ. The logs record the manager saying that he and the trailers wouldn't arrive until 4 p.m. on Sunday, June 30, and the video that was recording air tankers working the fire for the U.S. Forest Service's Aerial Firefighting Use and Effectiveness (AFUE) study shows the trailers approaching the fire at 3:38 p.m. on Sunday.
12. Karels et al., "Yarnell Hill Fire."
13. Wildland Fire Associates, "Granite Mountain IHC Entrapment and Burnover Investigation."
14. Ibid.
15. Karels et al., "Yarnell Hill Fire."
16. Ibid.
17. Ibid.
18. Holly Neill and John N. Maclean, "Discoveries in Yarnell Hill Fire Recordings Provide New Information About Location of Eric Marsh," Wildfire Today, January 19, 2014, http://wildfiretoday.com/2014/01/19/discoveries-in-yarnell-hill-fire-recordings-provide-new-information-about-location-of-eric-marsh/.
19. Christopher MacKenzie, "Granite Mountain Hotshots Last Video," YouTube, https://www.youtube.com/watch?v=omfw_Unt_VQ.
20. Karels et al., "Yarnell Hill Fire."

31. TRIGGER POINTS

1. Jim Karels et al., "Yarnell Hill Fire, June 30, 2013: Serious Accident Investigation Report" (State of Arizona, September 23, 2013), http://www.iawfonline.org/Yarnell_Hill_Fire_report.pdf.
2. Anne Ryman and Sean Holstege, "Yarnell Evacuation Flawed and Chaotic, Experts Say: Residents Describe Harrowing Ordeal," *Arizona Republic*, November 16, 2013, http://archive.azcentral.com/news/arizona/articles/20131116yarnell-fire-evacuations-chaotic-flawed.html.
3. Ibid.
4. Karels et al., "Yarnell Hill Fire."

32. NINETEEN

1. Jim Karels et al., "Yarnell Hill Fire, June 30, 2013: Serious Accident Investigation Report" (State of Arizona, September 23, 2013), http://www.iawfonline.org/Yarnell_Hill_Fire_report.pdf.

2. Holly Neill and John N. Maclean, "Discoveries in Yarnell Hill Fire Recordings Provide New Information About Location of Eric Marsh," Wildfire Today, January 19, 2014, http://wildfiretoday.com/2014/01/19/discoveries-in-yarnell-hill -fire-recordings-provide-new-information-about-location-of-eric-marsh/.
3. Karels et al., "Yarnell Hill Fire."
4. Ibid.
5. Statements by members of the Blue Ridge Hotshots (redacted), including Brian Frisby and Rogers Trueheart Brown, July 1, 2013, in Wildland Fire Associates, prepared for Arizona Division of Occupational Safety and Health, "Granite Mountain IHC Entrapment and Burnover Investigation: Yarnell Hill Fire — June 30, 2013" (November 2013), http://www.iawfonline.org/WildlandFire AssociatesReportYARNELL.pdf.

33. BLOWBACK

1. Russ Wiles and Robert Anglen, "Over $13 Mil Raised After Yarnell Tragedy: Big and Small, Donations Pour In for Hotshots' Families," *Arizona Republic,* December 24, 2013, http://archive.azcentral.com/business/news /articles/20131207yarnell-tragedy-millions-raised.html.
2. Pete Wertheim, "Prescott Examines Fiscal Impact of Wildfire Tragedy, Benefits, Hotshot Rebuilding," *Prescott eNews,* August 14, 2013, http://www.prescott enews.com/index.php/news/current-news/item/22104-prescott-examines -fiscal-impact-of-wildfire-tragedy-benefits-hotshot-rebuilding.
3. Carter Evans, "Ariz. Hotshot Widow Juliann Ashcraft Fighting for Denied Benefits," CBS News, August 5, 2013, http://www.cbsnews.com/news/ariz -hotshot-widow-juliann-ashcraft-fighting-for-denied-benefits/.
4. Tamara Sone, "City Won't Budge on Ashcraft Status: Widow Takes Her Plea to the Courthouse Plaza," *Prescott Daily Courier,* August 8, 2013, https://www .dcourier.com/news/2013/aug/08/city-wont-budge-on-ashcraft-status-widow -takes-he/.
5. "Minutes of the Special Voting Meeting of the Prescott City Council," Prescott, AZ, March 10, 2015, http://prescottaz.iqm2.com/Citizens/FileOpen.aspx?Type =12&ID=1160&Inline=True.
6. Felicia Fonseca, "Yarnell Hill Homeowners Suing State Alleging Mismanagement," Associated Press, June 25, 2014, http://www.firehouse.com /news/11534335/yarnell-homeowners-sue-arizona-saying-mismanagement -at-fault.
7. Jim Karels et al., "Yarnell Hill Fire, June 30, 2013: Serious Accident Investigation Report" (State of Arizona, September 23, 2013), http://www.iawfonline .org/Yarnell_Hill_Fire_report.pdf.
8. Wildland Fire Associates, prepared for Arizona Division of Occupational Safety and Health, "Granite Mountain IHC Entrapment and Burnover Investigation: Yarnell Hill Fire — June 30, 2013" (November 2013), http://www.iawfonline .org/WildlandFireAssociatesReportYARNELL.pdf.

9. J. J. Hensley and Yvonne Wingett Sanchez, "Feds Blocked Key Interviews in State Inquiry," *Arizona Republic,* December 6, 2013, http://archive.azcentral .com/news/arizona/articles/20131205feds-blocked-key-interviews-state-in quiry.html.

10. Bill Gabbert, "Forest Service's Explanation for Their Refusal to Fully Cooperate with Yarnell Hill Fire Investigations," Wildfire Today, December 9, 2013, http://wildfiretoday.com/2013/12/09/forest-services-explanation-for-their -refusal-to-fully-cooperate-with-yarnell-hill-fire-investigations/.

11. Holly Neill and John N. Maclean, "Discoveries in Yarnell Hill Fire Recordings Provide New Information About Location of Eric Marsh," Wildfire Today, January 19, 2014, http://wildfiretoday.com/2014/01/19/discoveries-in-yarnell-hill -fire-recordings-provide-new-information-about-location-of-eric-marsh/.

12. Yvonne Wingett Sanchez and J. J. Hensley, "Hotshot Team's Future in Question: Amid Vows to Rebuild, Logistical Hurdles Abound," *Arizona Republic,* August 17, 2013, http://archive.azcentral.com/news/arizona/articles/20130813hotshot -team-future-question.html.

13. Bill Gabbert, "Ironwood Hotshots to Be Disbanded," Wildfire Today, March 14, 2014, http://wildfiretoday.com/2014/03/04/ironwood-hotshots-to-be-dis banded/.

14. Lynne LaMaster, "Former Chief Darrell Willis Says ICMA Report on the Wildland Division Tampered With," *Prescott eNews,* March 17, 2015, http://www .prescottenews.com/news/current-news/item/25136-former-chief-darrell-wil lis-says-icma-report-on-the-wildland-division-tampered-with.

15. Robert Anglen, Dennis Wagner, and Yvonne Wingett Sanchez, "Yarnell Fire: New Account of Hotshot Deaths," *Arizona Republic,* April 3, 2015, http://www .azcentral.com/story/news/arizona/investigations/2015/04/04/yarnell-fire -new-account-hotshot-deaths/25284535/.

16. Cindy Barks, "Nearly 2 Years After Hotshot Tragedy, New Account Emerges," *Prescott Daily Courier,* April 9, 2015, http://www.dcourier.com/news/2015/apr /09/nearly-2-years-after-hotshot-tragedy-new-account-/.

17. Yvonne Wingett Sanchez, "Yarnell Hill Fire Lawsuits Settle for $670,000, Reforms," Azcentral.com, June 29, 2015, last updated June 30, 2015, http://www .azcentral.com/story/news/local/arizona/2015/06/29/yarnell-hill-fire-law suits-settlement-press-conference/29463359/.

18. "Yarnell Hill Fire Settlement Press Conference," Fox 10, Phoenix, June 29, 2015, https://www.youtube.com/watch?v=K7GASJl5T6Q.

19. "Mac and Gaydos Talk with Brendan McDonough," Mac & Gaydos, KTAR News, May 2, 2016, http://ktar.com/category/podcast_player/?a=321314&sid=1 002&n=Mac+%26+Gaydos.

EPILOGUE

1. Bill Gabbert, "Victims of Escaped Prescribed Fire in Colorado Receiving Set-

tlement Checks," Wildfire Today, July 29, 2014, http://wildfiretoday.com/tag/lower-north-fork-fire/.

2. U.S. Department of Agriculture, "Agriculture Secretary Tom Vilsack Announces Action to Combat Insects and Diseases That Weaken Forests, Increase Fire Risk," news release, May 20, 2014, https://www.fs.fed.us/news/releases/agriculture-secretary-tom-vilsack-announces-action-combat-insects-and-diseases-weaken.

3. "Senate Introduces the Bipartisan Wildfire Disaster Funding Act," Wilderness Society, January 22, 2015, http://wilderness.org/press-release/senate-introduces-bipartisan-wildfire-disaster-funding-act.

INDEX